The Green Screen Makerspace Project Book

Dec 2017

About the Author

Todd Burleson is a teacher-maker-librarian in Winnetka, Illinois. He was chosen as the 2016 *School Library Journal*'s School Librarian of the Year, and he has had the good fortune to present and speak throughout the world. He is passionate about finding a balance between books and bytes and helping children become real world problem solvers.

The Green Screen Makerspace Project Book

Todd Burleson

New York   Chicago   San Francisco   Athens   London   Madrid
Mexico City   Milan   New Delhi   Singapore   Sydney   Toronto

Library of Congress Control Number: 2017953566

The Green Screen Makerspace Project Book

1 2 3 4 5 6 7 8 9   LOV   22 21 20 19 18 17

ISBN   978-1-260-01995-7
MHID   1-260-01995-0

This book is printed on acid-free paper.

Sponsoring Editor Michael McCabe	**Copy Editor** James Madru
Editorial Supervisor Donna M. Martone	**Proofreader** Alison Shurtz
Production Supervisor Lynn M. Messina	**Indexer** Claire Splan
Acquisitions Coordinator Lauren Rogers	**Art Director, Cover** Jeff Weeks
Project Manager Patricia Wallenburg, TypeWriting	**Composition** TypeWriting

Contents

Foreword

I MET TODD BURLESON at the EdTech Summit in Boston in 2015. I was demonstrating our newly released Green Screen by Do Ink app for the iPad and iPhone. Todd is a former third and fourth grade classroom teacher who a year after our meeting was named the 2016 School Librarian of the Year. This award is given by the *School Library Journal* and sponsored by Scholastic Library Publishing to honor K–12 "school library professionals for outstanding achievement and the exemplary use of 21st-century tools to engage students toward fostering multiple literacies." Todd's book, *The Green Screen Makerspace Project Book*, reflects his passion for learning and is a wonderful introduction to green screening. It's a compilation of authentic classroom projects that will help you get started with green screening and spur your imagination as to all the ways green screening can be used to engage students.

You know what a *green screen effect* is, right? It's used in the movies to make it look like the actors have landed on an alien planet, and it's used on television to make it look like your local news announcer is standing in front of a weather map. The green screen effect works by combining images from multiple sources into a single video or photo. The genesis of our Green Screen by Do Ink app came from a teacher at EdCamp Boston in 2013 who mentioned a need for an easy-to-use green screen app. She wanted to teach a weather segment and needed the app to be easy enough for her second grade students to use. As cofounder of Do Ink, an educational app company, I recognized the need to involve teachers early in our development process and to test the results in classrooms. This input was invaluable. We launched our Green Screen by Do Ink app, and two weeks after its launch, Apple selected it as a "Best New App in Education." Fast forward four years, and we were amazed at the endless ways green screening is used in classrooms throughout the world.

Green screening used to be difficult and time-consuming and required expensive equipment. Now it can be done without fancy studio setups using inexpensive software on tablets, phones, and computers. Green fabric, green tablecloths, green-painted pizza boxes and trifolds, green file folders, and even green Play-Doh have been used as green screens to great success in the classroom. The popularity of green screens in the classroom is due to the fact that they immerse students in their learning. One minute they can be in China on the Great Wall and the next, exploring the moon.

This book is a wonderful introduction to using green screens in education and illustrates with sample projects ways it can be used in twenty-first-century classrooms. I hope this book will inspire you to try green screening because it is a fun and innovative way for students to be creative and to communicate their learning because it promotes student collaboration, critical thinking, and problem solving.

Karen Miller
Cofounder of Do Ink

Acknowledgments

WRITING A BOOK CAN BE both a communal and solitary undertaking. To that end I would like to thank some of the folks who have been a part of this process. I couldn't have even begun this process without Colleen and Aaron Graves. They recommended me to my editor Michael McCabe because of my experience working in our school television studio and with a variety of green screen technology. I am grateful to them nudging me to create this book. Michael McCabe, my editor, has been incredibly patient and supportive of me as a new author. Thank you to the phenomenal educators who helped generate ideas for this book. Some of them didn't make it into the book with specific projects, but their shared wisdom was essential. Thank you to each of them: Billy Spicer, April Wathen, Madonna Marks, Sarah Guillen, and Sherry Gick. Thank you to the administrators in my district who have encouraged and supported my work, specifically to principals Maureen Cheever, Daniel Ryan and Beth Carmody, tech director Maureen Miller, and Superintendent Dr. Trisha Kocanda. Most importantly, I want to thank my family for putting up with me spending countless hours away from them while writing this book.

CHAPTER 1

Introduction

As EDUCATORS, we are always looking for ways to inspire creativity and wonder in our students. Green screen technology allows us to help our students experience "magic." Imagine being able to fly, shoot lasers from your fingertips, hold the Eiffel Tower in your hand, report from the International Space Station, or deliver the weather from inside a volcano. With green screen technology, your students can do all of these and much more (Figure 1-1).

It is no longer necessary to have a Hollywood budget or professional-grade lighting and camera equipment to achieve high-quality special effects. With some very basic materials

Figure 1-1   A student takes flight!

and a little technology, it is possible to create this type of magic (Figure 1-2).

Putting these tools in the hands of students allows them to extend their creativity and produce unbelievable presentations, projects, and videos. Where it would have cost millions of dollars and a small army of assistants, students are now only limited by their imaginations. Because more and more of our world is experienced in the "digital realm," increasing the

quality of presentations increases the likelihood that viewers will invest the time to absorb them (Figure 1-3).

This book is designed to empower both students and teachers. In it I'll review how early filmmakers created mesmerizing special effects and show how these technologies evolved over the last century. I'll examine and compare different methods and tools. Last, I will walk you through two dozen projects step by step. Hopefully, these projects will inspire dozens more. Once you teach your students these skills, watch out and be prepared to be amazed at what they create (Figure 1-4)!

This book is first and foremost for educators, not cinematographers. Professionals might cringe at the quality of some of the projects students create. The point, though, is that the technology has advanced to the point that a kindergartner *can* make a green screen video. While they might not win any Academy Awards, these projects will inspire learners of all ages and will lead them to create and produce projects that will open their worlds (Figure 1-5).

Figure 1-2   *Hockey Magazine* cover in process.

Figure 1-3   Standing in the hand of the Big Friendly Giant.

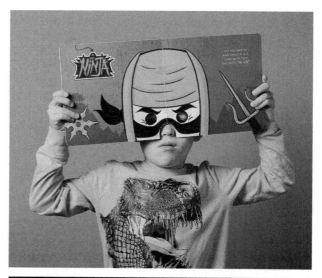

Figure 1-4   This ninja is ready for a new background!

Figure 1-5 A look inside the WGST (The World's Greatest Student Television) studio during a morning broadcast.

CHAPTER 2

Evolution of Green Screen Technology

IT'S IMPORTANT TO REMEMBER that moving pictures, when they were first introduced, were in and of themselves quite magical. For the first time ever, audiences saw moving images on a screen. It was probably a lot like the experience many of us have when we experience virtual reality for the first time. What appeared to be lifelike movements of actors were in fact merely successive individual photographs taken and played back 24 frames per second. The images were created on light-sensitive silver-covered material exposed to very bright light and then "developed" in a chemical bath. Over a hundred years later this process that still evokes a sense of magic in those who experience it in the darkroom. Early films were short, black and white, and silent. They typically showed just one scene. It wasn't until 1927 that sound would be able to be recorded and played in sequence with the film.

Moving pictures were an immediate success, and audiences clamored for more. In 1898, Georges Méliès amazed viewers of his film *The Four Heads* (Figure 2-1). It is one of the earliest examples of special effects and must have absolutely mesmerized his audiences. Because the film was in black and white, he used glass plates painted black to "mask" certain parts of the scene, notably his head. In *The Four Heads*, Méliès filmed the scene with the glass plate over his head to give the illusion that he had taken

Figure 2-1 Georges Méliès (1861-1938), French filmmaker and cinematographer (https://upload.wikimedia.org/wikipedia/commons/6/63/George_Melies.jpg)

his head "off." He then rewound the film and re-exposed the film with his head on the table. He rewound and re-exposed the film several more times with his head in different places. The resulting double, triple, and quadruple exposures gave the illusion that his head had multiplied! It shocked and amused his audiences. What amazes me to this day is that he did all this "in camera," not being sure that he had the "shot" until after he developed his film. He was a genius! (Figure 2-2).

Figure 2-2   *The Four Troublesome Heads*
(https://en.wikipedia.org/wiki/The_
Four_Troublesome_Heads#/media/
File:M%C3%A9li%C3%A8s,_Un_homme_
de_t%C3%AAtes_(Star_Film_167_1898).jpg)

As film technology evolved, so did special effects techniques. Soon a technique called a *traveling matte* allowed actors to move in front of objects that were not filmed in the original scene.

In 1918, Frank Williams patented his technique, which he later used in *The Invisible Man* (Figure 2-3). By dressing his actor in a head-to-toe black velvet suit against a black background, he was able to make it appear as if the actor was indeed invisible. Even today the effect is humorous and quite believable.

In 1925, a new technique was developed called *blue screen* or the *Dunning process* by C. Dodge Dunning. Dunning's process worked by capitalizing on the qualities of light. The subject was lit with bright yellow light against a blue background. Through processing the film with a variety of dyes and filters, the different colors could be separated and printed onto film creating traveling mattes. One of the earliest examples of this technique is *King Kong* from 1933 (Figure 2-4).

The Thief of Baghdad, which was produced in 1940, is an excellent example of the blue screen technique. A genie is seen to escape from a bottle. This film went on to win the Academy

Figure 2-3   *The Invisible Man Film*, 1933
(https://en.wikipedia.org/wiki/The_Invisible_
Man_(film)#/media/File:The-Invisible-Man.jpg)

Award for special effects. By today's standards, the effects seem pretty gimmicky, but they were mind-blowing at the time. The blue screen technique was used extensively in Hollywood but was incredibly meticulous. The technique had to be done frame by frame through several passes with myriad filters and development tools. Different techniques using various frequencies of light and filters continued to develop through this time as well.

Walt Disney developed a technique for color film that was seen in the classic film *Mary Poppins* from 1964 (Figure 2-5). It required filming the scene with very bright sodium light, which produced a distinctly orange cast on a white background. To make this technique work, it required an incredibly advanced printing

Figure 2-4 *King Kong*, 1933
(https://en.wikipedia.org/wiki/King_Kong_
(1933_film)#/media/File:Kingkong33
newposter.jpg)

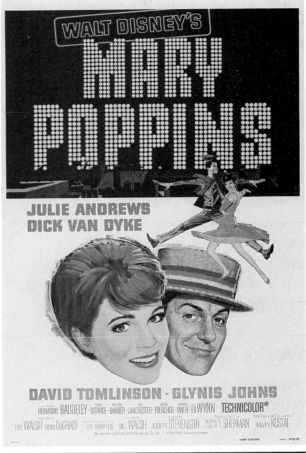

Figure 2-5 *Mary Poppins*, 1964
(https://en.wikipedia.org/wiki/Mary_Poppins_
(film)#/media/File:Marypoppins.jpg)

technique and a literal "one of a kind prism" that split the light.

With the development of microcomputers in the 1980s, the process could be automated and sped up exponentially. Richard Edlund won an Academy Award for the technology he developed that controlled the optical printer more efficiently in the making of one of my favorite films of all time, *The Empire Strikes Back*. The more I learn about how intensive the process of creating films such as *The Empire Strikes Back*, the more I am amazed that such films ever were created. Further advances in computer technology have allowed for precipitous growth in the variety and complexity of visual effects.

So, when and why did filmmakers start using green screens? The color green is used as a background when compositing scenes for two reasons. First, it is very different from the typical reddish color of human skin, and thus actors can be separated from the background more efficiently. If you look at a color wheel, red is on the opposite side of green. Second, film and digital camera sensors, like the human eye, are more sensitive to green light. As digital cameras have become more and more prevalent, so has the green screen technique. Filmmakers today still use both blue and green screens when compositing depending on the coloration of the scene, the character's costume, and different lighting situations.

How exactly does the green screen, also known as the *Chroma key*, work? Essentially, the technique eliminates or isolates the narrow band of color contained in the green screen and replaces it with anything you can imagine (Figure 2-6). The screen does not have to be green; it can also be blue. Blue is also far from the color of human skin and is in some cases preferable to green. This is particularly true if the costume the actor is wearing contains green or the setting is at night.

Today, green screen technology has advanced to the point that it does not require expensive lighting or special materials. In fact, you and your students can create projects with stunning results with just a few inexpensive tools.

Figure 2-6   This split screen shows how the green screen can be replaced.

Works Cited

"Double Feature." *Wikipedia*. Wikimedia Foundation, June 13, 2017. Accessed June 19, 2017.

"Georges Méliès." *Wikipedia*. Wikimedia Foundation, June 15, 2017. Accessed June 19, 2017.

"The Invisible Man (film)." *Wikipedia*. Wikimedia Foundation, June 19, 2017. Accessed June 19, 2017.

"King Kong (1933 Film)." *Wikipedia*. Wikimedia Foundation, June 18, 2017. Accessed June 19, 2017.

"Visual and Special Effects Film Milestones." An Award-Winning, Unique Resource of Film Reference Material for Film Buffs and Others, with Reviews of Classic American-Hollywood Films, Academy Awards History, Film Posters. Accessed June 17, 2017.

CHAPTER 3

Green Screen Materials

GREEN SCREENS CAN RANGE in cost and complexity from a dollar to hundreds of thousands of dollars. Because this book is geared toward educators who are always on a tight budget, I will give a broad range of materials and installations for reference and let readers decide for themselves what is best for their applications and budgets. My best advice is to start small. Soon enough your students will be begging for more!

I mentioned earlier that the term *green screen* has become synonymous with the special effect of removing a background and replacing it with something else. The *screen* itself does not have to be green. It can also be blue; these two colors are not present in human skin tones. Digital cameras are highly sensitive to the wavelengths of these two colors, making them easier to remove. Green is a more luminescent color than blue, so it takes less light than a blue screen. For the sake of simplicity, we'll focus on green screen applications, but know that the color blue can be used as well (Figure 3-1).

A true green screen has a very specific color. Pantone, the world standard for color, specifies specific colors with its Pantone Matching System (PMS). According to Pantone, the pure green screen color is number 354C.

One of the most effective and professional ways of creating a green screen is to paint a smooth wall. There are a number of brands

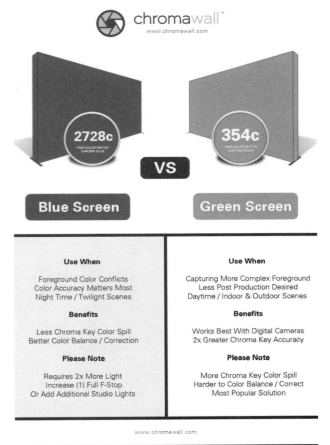

Figure 3-1 When to use a blue or a green screen.

of Chroma key–specific paint. The industry standard is Rosco Chroma Key Paint. It is considerably more expensive than consumer-grade latex paint, but the professionals stand by the results (Figure 3-2).

Figure 3-2 A green screen studio made by painting both the wall and floor Chroma Key Green.

For less than half the price of this paint, you can get a gallon of flat latex paint from your do-it-yourself (DIY) store and achieve nearly identical results. The specific name or brand of the paint may vary, but one brand with which I have had excellent results is Behr Deep Base Number 1300 interior flat latex and the name of the color is Sparkling Apple S-G-430. The downside of a green wall is that it is not transportable. One fun advantage of green screen paint is that you can paint other items, such as tables, boxes, and so on. These can be helpful to support smaller items or even to place your subject.

I've seen some incredible installations that make use of what is called a *cyclorama*. These curved surfaces create the illusion of infinite space. A cyclorama is often called an *infinity background* or a *cyc* for short. Cycloramas range from tabletop applications to entire studio installations (Figures 3-3 and 3-4).

One of the most impressive transportable green screen setups I've come across is the ChromaWall Retractable Green Screen. This self-standing screen is 59 inches wide and 78.5 inches tall (Figure 3-5). The manufacturer sells a smaller retractable version as well (Figure 3-6).

Figure 3-3 A cyclorama or an infinity wall installation.

Figure 3-4 A professional cyclorama by ProCyc.

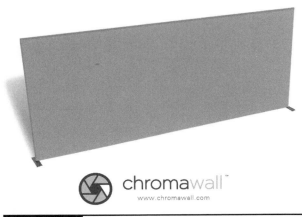

Figure3-5 A ChromaWall collapsible wide studio backdrop.

Figure 3-6 A ChromaWall collapsible narrow studio backdrop.

A much less costly setup is the 60- × 72-inch Reversible Chroma Blue and Green Background Kit. This kit comes with a stand, and the fabric background is collapsible for easy transport. This kit costs approximately $80 online. My school has two of these kits in our library, and we often do not even need to use the stand; we

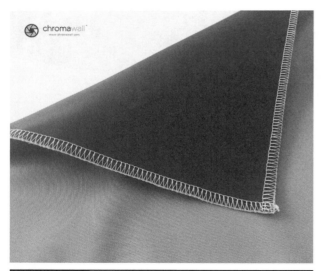

Figure 3-7 A reversible blue/green screen.

just lean the background up against a flat surface (Figure 3-7).

Green fabric is another alternative. Specialty fabric can be purchased at a premium. An economical option is to purchase several yards of heavy cotton fabric at your hobby store and sew the panels together to achieve the width you need for your application. In my school television studio, we use a simple backdrop stand to support our green screen. We sewed horizontal "pockets" at the top and bottom of the cotton panels. Horizontal poles are inserted into these pockets, and the weight of the poles holds the fabric tight (Figure 3-8).

Green screens need not always be large enough to fit one or more standing figures. Many projects in this book are accomplished with much smaller scale green screens. Most are simple and easy to make.

Just about any smooth, nonshiny surface can be used as a green screen. My top 10 green screen ideas that cost under $5 are listed in Figure 3-9. These and more green screen ideas are shown in Figures 3-10 through 3-15.

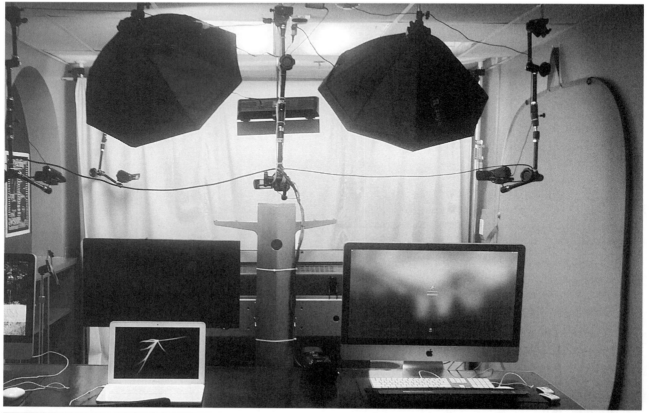

Figure 3-8 The WGST (World's Greatest Student Television) studio at Hubbard Woods Elementary School.

Figure 3-9 Ten easy ways to get started with green screen illustrated by Larissa Aradj.

Figure 3-10 Green foam, paper plates, a dollar-store tablecloth, and poster paint can be used in a variety of ways to make green screens.

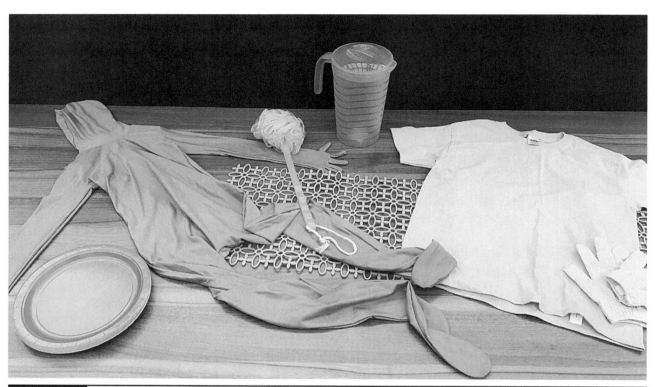

Figure 3-11 Green "morph" suit, green T-shirts, and other items can be fun to experiment with when exploring green screen.

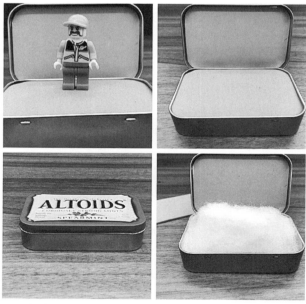

Figure 3-12 A mini green screen made from an Altoids box enables spontaneous creativity.

Figure 3-13 Experiment with painting masks, foam heads, and more with green paint to create unique green screen creations.

Figure 3-14 A green file folder can be used as a simple green screen.

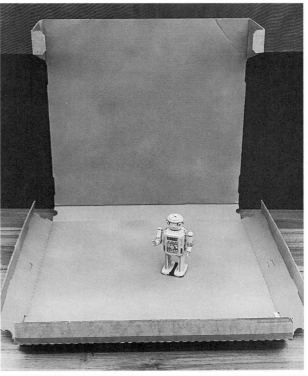

Figure 3-15 A clean pizza box painted green is a simple green screen that doubles as a place to store green screen props.

Do you have an interactive whiteboard in your classroom? Google Pantone 354C, and you can turn a whiteboard into a green screen! You can use your tablet as a smaller green screen, as I do in Project 7 (X-Ray Machine) by doing the same thing. Do you have an old projector screen? Use spray adhesive to mount some green cotton fabric or felt to it, and you have a retractable green screen.

Ryan O'Donnell has created kits that teachers can check out to do green screen projects in their rooms. Hardware aside, these kits can be made for a fraction of the cost of many found online. In his blog post about making these kits, O'Donnell reflected on the kit he had purchased from Amazon:

1. I've never used the lights it came with . . . too bulky and just not really needed. The lighting in most classrooms is just fine.

2. It's a bit large. Both lugging it around and setting up the stands is not super quick.

3. There is only one of them. When trying to have student groups use them, there was always a line of kids waiting their turn.

These simple kits contain seven items (Figure 3-16):

1. **iPad mini 2.** At $269, this is the tool. You can use other iPads, but this cost and size are best suited for students. What makes it perfect is that on just this one device you can shoot, edit, and share your green screen creations. As of now, there is no other android tablet and green screen program that can rival it.

2. **Green Screen by Do Ink app.** Several other apps are available, but this is the most user friendly and robust in the app store. Plus, the manufacturer is continually updating the program with new and improved features.

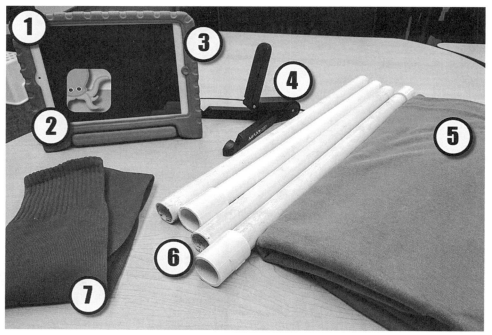

Figure 3-16 Items in a portable green screen studio created by Ryan O'Donnell.

3. **iPad case.**

4. **Max Table Stand.** You will need to keep the iPad as steady as possible to create a realistic illusion. You can try to use a tripod, but I find a stand such as this one from Max Cases to be small and easy to use.

5. **Green fabric.** For each of my kits, I used 2 yards of fabric. I folded one side over a few inches and sewed a seam. It is this seam through which you will feed the PVC pipe.

6. **PVC pipe and coupler.** I went with a ¾-inch pipe, small but still strong enough to hold up the weight of the fabric.

7. **Green socks.** Not needed but fun addition to put on hands to help make things "appear" from off screen.

O'Donnell suggests using couplers to make the overall length of the PVC tubes shorter so that they can fit inside a single travel bag. He also recommends that you make several so that groups of students can work at the same time.

Jennifer Leban, an incredibly creative educator, came up with a "wearable green screen" with her students. This prototype uses recycled hangers, duct tape, and green foam core to create a green screen that is perfect for the webcam she uses with WeVideo. What I love most about the blog post where she details this creation is the way she is constantly thinking about improving the design. She freely admits that it isn't perfect. She models the importance of the design cycle for her students (Figure 3-17 through 3-19).

Wendy Garland, a librarian from Massachusetts, discovered that the green straws that Starbucks uses fit perfectly into the hole on a Shopkins toy. These make fun green screen "puppets" in front of a simple background (Figure 3-20).

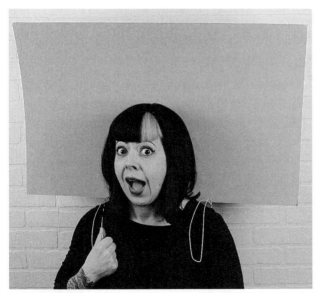

Figure 3-17 Jen Leban's wearable green screen in action.

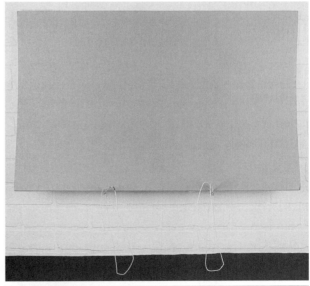

Figure 3-18 A view of the shoulder braces of the wearable green screen.

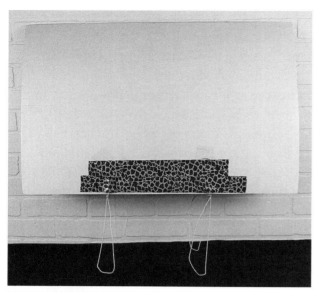

Figure 3-19 The back view showing how the braces are anchored using duct tape.

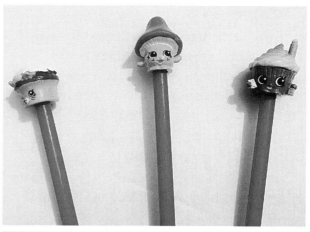

Figure 3-20 Shopkins have a hole that perfectly fits Starbucks green straws and make fun green screen props.

Another way to help make objects move on a green background is to attach very light fishing line to your puppet or object. The line pretty much disappears when an image replaces the green screen.

No matter what you use to create your green screen, you'll want to have some Chroma key fabric on hand. This is particularly useful to extend your green screen by covering the ground, a milk crate, a table, or whatever you want to use in your scene.

Green foam rubber tiles work well to expand a pop-up green screen background to the floor. An added benefit is that the foam rubber is soft and easy to clean.

Want an invisibility cloak? Drape that extra bit of green fabric over your actor, and you have one! Looking to capture some photos of your students performing gravity-defying skateboard tricks? Lay the green screen on the ground, and use a ladder to shoot from high above.

Janelle Van Dop used green screen paint in a totally new way. She recorded her students painting a white background with Chroma key green paint to reveal an image that appeared to be underneath. This is such a simple but creative way to use the technology.

Dress your model in a Chroma green morph suit and have him or her move objects within the scene. Or have your model wear the suit except for his or her head, arm, leg, and so on. Try it with Chroma key green gloves or a mask. Anything green becomes invisible.

Have an old treadmill? Paint it green, and you have a perfect tool for realistic movement. Rosco has a blog post that will walk you through the process step by step (www.rosco.com/spectrum/index.php/2015/12/how-to-paint-a-greenscreen-treadmill/).

A solid support for the device capturing the photos or videos enhances the entire project and protects your tablet. There are dozens of tablet cases and stands available. Here are a couple of my favorites:

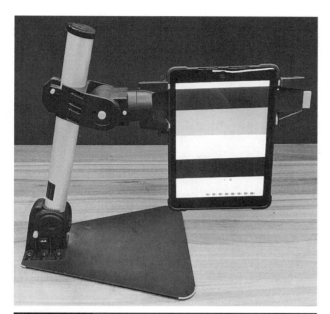

Figure 3-21   The Juststand version 2
(https://ipaddocumentcamera.com/pages/
justand-v2)

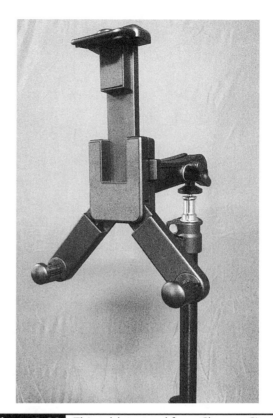

Figure 3-22   This tablet stand from Charger City
tablet stand can be used with a
camera tripod.
(http://www.chargercity.com/HDX2-RM8-Video
-Record-Periscope-Tablet-Tripod-Mount-with
-Dual-360-Swivel-Adjust-Joint-for-7-12-inch
-Tablets-like-Apple-iPad-Pro-Air-Mini-Galaxy
-Tab-S2-A-Note-Surface-Pro-Slate-c-713-p-755
.html)

- The Juststand is a fantastic iPad stand. This stand is incredibly adjustable and very sturdy (Figure 3-21).

- If you have an old tripod hanging around, the tablet stand from Charger City will quickly screw onto the tripod head (Figure 3-22).

Green screen compositing would be pointless without some quality background content. It has never been easier to find and use appropriate content. One of the most effective ways of finding images is to use an image site such as www.Photosforclass.com, www.Pixabay.com, or www.pics4learning.com. One very useful tool to teach students is how to properly look for images using Google Images. Simply search Google Images. Then click on the Tools tab. Then click on the Usage Rights tab. Select Labeled for Reuse. I also have students add another variable; I have them select the size of the image. Low-resolution images look terrible in final digital projects.

Of course, costumes and props are infinitely valuable for green screen projects. Visit your local costume shop the day after Halloween to score big and save a ton. I collect these items and just about anything else that's green and keep them near my portable green screens for whenever inspiration strikes (Figure 3-23).

Figure 3-23   It's a great idea to keep props of all sorts on hand for when creativity strikes!

CHAPTER 4

Green Screen Lighting

WHILE THIS BOOK ISN'T GEARED toward cinematographers, it's important to understand the science and technology behind green screen. Lighting is important. After all, the very concept of green screen replacement, or *compositing*, is based on the way light waves are absorbed by matter and how a camera sensor processes that information.

The single biggest change I've noticed with green screens in the decade and a half that I've been using them is lighting. When I began using green screens, they required an immense amount of very bright white light. The halogen bulbs were extremely hot and costly to replace. This was not only uncomfortable in my tiny studio but also potentially dangerous. I had up to eight lights placed on stands throughout the studio. If a student touched one of those lamps, he or she could suffer a severe burn. I was also always concerned about the potential for fire. The Chroma key hardware and software needed this light in order to separate and remove the background. My setup looked much like the one in Figure 4-1.

Today, there are many more alternatives. Here are a few examples of permanent studio setups.

One school I visited in Colorado had an amazing setup. It included a cyclorama (or cyc) that gives the illusion of an infinite set (Figure 4-2). This figure illustrates a few elements of a dream studio setup. First of all, the lights

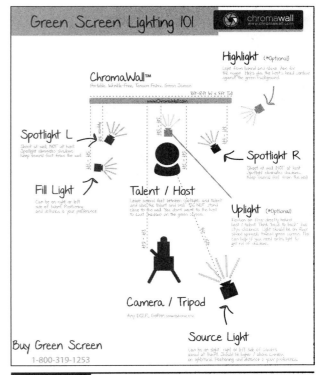

Figure 4-1 An "ideal" lighting situation for a green screen studio.

mounted on the ceiling eliminate the wear and tear (as well as the hassle) of placing light stands. The three lights closest to the backdrop provide smooth, even lighting of the green cyclorama. The two large lights off to the sides are called *soft boxes*. They produce an even "soft" light that illuminates the actors evenly. When taken together, these two different lighting *planes* produce no shadows on the

Figure 4-2 A cyclorama complete with ceiling mounted LED lights.

background. This makes compositing complex backgrounds simple. Last, the *green box* can be used to simulate just about any type of element in postproduction. It could be a tree, a car, or a giant monster's hand! It can also be used to allow the actors to "fly" by having them lie down on the box.

When my school relocated its television studio to a place with a much smaller footprint, we decided to mount all our lights on the ceiling. You can also see that we mounted our cameras on arms that extend down from the ceiling as well (Figure 4-3).

By putting all of these fixtures on the ceiling, we got the feeling of more space, and we avoided tripping over cords and stands. This setup works perfectly for our daily news broadcast but isn't ideal for spur-of-the-moment projects because of the fixed camera locations. For spur-of-the-

moment projects, we have multiple portable green screens that we can drag wherever they are needed (Figure 4-4).

In this figure you can see a portable green screen balanced against a wall. There is zero natural light in the hallway where this is being recorded, so we are using a battery-operated light-emitting diode (LED) light that screws on to our Padcaster frame. The Padcaster is a rugged frame for the iPad. The kit comes with this LED light, a microphone, and a sturdy tripod. When paired with the Green Screen by Do Ink app, it's a mobile television production studio that rivals the $10,000 we began our television studio with back in 2001.

To get the highest-quality image, the background and the actors need to be lit evenly and effectively. Any wrinkles or "hot spots" in the background will cause problems when you

Figure 4-3 The permanent green screen setup in our school television studio.

Figure 4-4 The Padcaster and portable green screen enable us to take our studio anywhere we need to.

try to replace the green screen. For this reason, I love pop-up green screens. They can be propped up just about anywhere and are inherently wrinkle-free. It is also essential to avoid shadows. Simply having your subject stand away from the background and then slightly tweaking your foreground lighting can accomplish this. In Figure 4-4, the student recording the segment *should* have had the actor step forward a bit to get rid of the shadow. You can see from the preview on the iPad screen that the software is remarkably forgiving.

To achieve *professional* results today, it is still advisable to have multiple lights, but LED or fluorescent bulbs now can power those light sources. The result is far cooler lights and less expensive replacements. These lights can also be placed on ceiling-mounted fixtures. If you have a

Figure 4-5   The Padcaster's LED light enables a quick and high-quality setup without having to use multiple lights and stands.

dedicated space for your green screen studio, this is an excellent setup.

If you don't have a permanent space for your green screen, you can still achieve high-quality results. Inexpensive lighting kits are readily available. They typically include two lights for the background and two for the subject. Very few teachers actually use them because they just take too long to set up and take up far too much space (Figure 4-5).

Many teachers place their green screens in a spot that receives even lighting, either from diffused window light or even from overhead lighting. My school just had all of our fluorescent ceiling lights replaced with LED fixtures. The quality of the light is much more pleasant and makes it easy to quickly capture green screen footage.

As soon as students discover how quickly and easily they can use Chroma key techniques, they begin pushing the concept. Just about anything green can become a green screen. I've loved doing stop-motion animation with my students. When one of them asked if we could use the green Lego base plates as a green screen, I had no idea. It turns out that you can! From then on, the stop-motion videos the students created reached an entirely new level of sophistication. They were now able to change angles and perspectives in ways we never had explored before. They began thinking critically about their sets and backgrounds. Even our Lego wall, which is made of blue base plates, has been used as a blue screen for puppet shows and stop-motion films.

Having multiple mini–green screens has the added bonus of allowing many more students to create at the same time. Take the base plate example just mentioned. I use two of these base plates taped to a bookstand to create an individual green screen "set" for my students. Now I can have 10 to 12 pairs of students working at the same time. I'm blessed with huge windows in my space, so the students place their sets toward the windows. I've seen teachers who don't have great lighting in their rooms use everything from clip-on LED lights to superpowerful flashlights diffused with a sheet of tissue paper.

CHAPTER 5

Production Software and Apps

GREEN SCREEN PRODUCTION SOFTWARE varies widely in complexity and cost. There are programs designed solely to extract an object from its background and others that serve a wide range of purposes. For perspective, I'll highlight the range but focus specifically on the tools that are most useful for educators.

Some instances call for the removal of a background in a still photograph. The world standard for photograph editing is Adobe Photoshop. This software is quite expensive, can take years to master, and is constantly evolving. One thing I have learned in studying Photoshop is that there are literally dozens of ways to do the same task. This is especially true for green screen photography, or *compositing*. There are powerful tools called *plug-ins* that can be purchased to speed up the process of photo and video editing. These plug-ins could be thought of as microprograms that focus on one specific function: removing a green background. For the average user, certainly for a student, they are quite costly. In contrast, the free iPad app called *Magic Eraser* allows users to do the same thing on tablets or phones in seconds. Not only is the powerful program free, but it also gives you the option of saving your composition as either a JPEG file on a white background or as a PNG file. A PNG file has a transparent background. This is particularly helpful when compositing because it allows you to layer the image easily.

With a tool such as Magic Eraser, it is possible to remove the background of an image that was not recorded in front of a solid green or blue background. This doesn't work for video, but for stills it is remarkable (Figure 5-1).

To edit and record green screen video, the first challenge is to define your platform: Mac, Windows, phone, tablet, or netbook. Each has a multitude of options available to it.

As schools move closer to one-to-one devices, the choice of device can have a significant impact on how these tools are used. Until recently, the choice of a laptop device came down to operating system: Windows versus Macintosh. Now a third choice has entered the market: netbooks powered by the Google Chrome operating system, commonly referred to as *Chromebooks*. The number of Chromebooks has grown exponentially primarily due to their low cost. As the number of these devices has ballooned, so has the number of applications that can be used with them.

One of the most popular cross-platform pieces of software is WeVideo. WeVideo can be used on any computing device: Mac, PC, Chromebook, iOS, android, and even a web browser. This diverse program offers two modes: storyboard and timeline. This makes it easy for students as young as kindergarten age to build their projects and is expansive enough for budding cinematographers. The program includes a basic

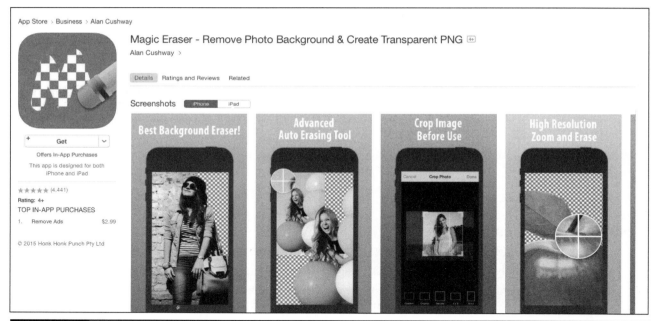

Figure 5-1 The Magic Eraser App

Chroma key editor. Chromebook web cameras are less than magnificent, but they do allow you to capture images and low-quality video. I recommend recording video on another device and then uploading it to the web via Google Drive or a similar web storage facility. The video can then be edited via WeVideo and shared (Figure 5-2).

WeVideo offers three levels of access: free, power, and unlimited. The free version is limited in most ways and does *not* include green screen or many of the advanced editing features.

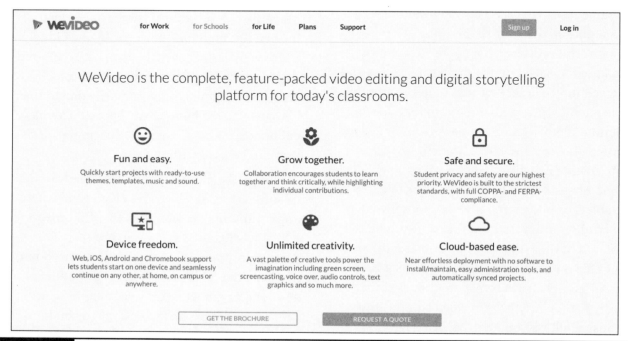

Figure 5-2 WeVideo is a powerful multi-platform video creation tool.

WeVideo allows users to sync their work across device and platform. Users can create a project at home on their phone and then complete it at school in their web browser simply by completing their work within their WeVideo account. WeVideo seamlessly integrates with Google apps, making collaboration and sharing simple. Another reason I especially like WeVideo is that it is a self-contained environment. Students can create green screen compositions and then layer in music, titles, special effects, and more all within the single WeVideo system. There is no need to export from one app and then import to another. For younger students, this simplifies the process significantly.

The simplest and most beautiful way for Apple users to create videos on their computers is iMovie. iMovie is included on every Macintosh as part of its operating system. This video editing and creation platform might lack a few of the advanced features available on professional editing programs such as Adobe Premiere and Final Cut Pro, but it is simple to use and exceedingly versatile. The green screen feature in iMovie is as easy as adding a layer and clicking on the green screen feature. I have used this program in my school news studio for years.

Tablet computers are now the most popular devices in schools. Their costs continue to decrease as their features and storage increase. It is possible now to purchase an Amazon Fire tablet for just over $50. As mentioned earlier, WeVideo is a universal tool and one of the best on android tablets. My preference for a tablet computer is the iPad.

The iPad is also my favorite tool for digital creation. The inherent challenge of its tiny built-in storage is nixed with add-ons such as Google Drive, Google Photos, Dropbox, and more. There are two green screen iPad apps that are designed for elementary-age users but are powerful enough for advanced users as well: Veescope Live and Green Screen by Do Ink. Veescope Live packs some fantastic features, including up to 12-megapixel photographs and 4-kilobyte-resolution videos. One of the features that stands out to me is the fact that you can use the Chroma key feature on a blue or green screen but also on any blank white wall. This means that you can "key out" simply without the need for a green screen; you just need a solid-color background. It also has some advanced features that let you fine-tune your lighting and create the highest-quality video and images. A final feature that sets it apart is the ability to add a variety of built-in layers such as a weather map and some basic virtual sets (Figure 5-3).

Figure 5-3   Veescope Live green screen app.

The app I typically recommend for educators and students who are just getting started with Chroma keying on iPads is the Green Screen by Do Ink app (Figure 5-4). The simple interface is easy for users of all ages to master. The feature that makes Do Ink stand out even further is its ability to combine up to three layers in a single video or photographic file. At first, this may not seem to be that big of a deal, but when you combine graphics or animations created in other apps such as Animate by Do Ink, you can expand the creative potential exponentially. Saving layered compositions and then reimporting them and adding further layers can expand this process infinitely. Most of the projects in this book were created with the Green Screen by Do Ink app, but with a little creativity, they could be done with other apps or tools as well.

Neither of these apps is designed for final video production. They are terrific at creating green screen effects, but tools such as iMovie for iOS and WeVideo can be used to combine multiple video segments and add sound effects, transitions, recorded audio, or a soundtrack.

TouchCast is a massively powerful iOS app designed to have students broadcasting in minutes. This fantastic tool is chock-full of excellent features, one of which is green screen. The feature that gives TouchCast its name is the interactivity of the multitude of layers that can be added. They call it *smart* video, and viewers can interact with or "touch" each of the elements on-screen. This powerful program includes every feature you might find in a broadcast studio from an on-screen teleprompter to advanced lower thirds. Users can get started with the wide array of built-in templates. The powerful green screen editor includes dozens of backgrounds and virtual sets. This app is perfect for schools that want to produce a daily news-type program on a budget. Users can also customize and develop their own templates as they gain experience. Amazingly, this program is free! It is robust enough to be used by leading companies around the world and will have your students creating engaging videos in minutes (Figure 5-5).

TouchCast shares its mission:

> We have a mission—to support education reform in public schools by giving students deeper learning experiences. We are all aware of the many challenges that the public education system is facing. At TouchCast, we have the opportunity to impact millions of

Figure 5-4   Green Screen by Do Ink.

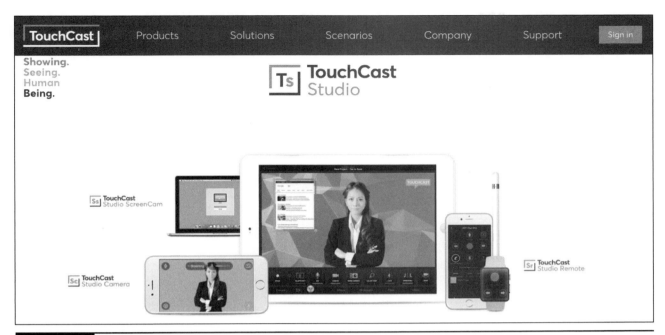

Figure 5-4   Touchcast Studio.

children that go to school every day. With the constant digital stimulation around us, engaging with this YouTube generation has never been more difficult. Young people today are drawn to the world of online broadcasting and wish to communicate through video with their local and global community. By supporting their natural communication habits, we are helping to create an environment in which they learn by doing. Hundreds of thousands of students create TouchCast Smart Videos in their classrooms every week—and they have fun doing it. We are thrilled to support students and teachers, who work so hard with so few resources. Our objective is to improve their lives and education—which is why we make TouchCast Studio available for free to anyone in the public school system. We are fortunate enough to be able to pursue our social mission while we continue to build a vibrant business serving the world's largest and most progressive companies.

This is an amazing time to be an educator! A nearly unlimited number of pathways exist for teachers and students to bring their imaginations to life. I hope that that this book serves as a springboard for those ideas and encourages creators to experiment and explore.

Works Cited

Coughlan, Sean. "Tablet Computers in '70% of Schools,'" *BBC News*, December 3, 2014. Accessed June 17, 2017.

"Our Social Mission." TouchCast, June 17, 2017.

Sunshine Open Solutions. "The 5 Types of Digital Image Files: TIFF, JPEG, GIF, PNG, and Raw Image Files and When to Use Each One." IvanExpert Mac Blog, September 29, 2011. Accessed June 17, 2017.

CHAPTER 6

Projects Step by Step

MOST OF THE PROJECTS in this book were created using the iPad. This is true for two reasons. First, I believe the iPad to be the easiest and most flexible tool for students to use to create media. Second, it also happens to be the device that my school district provides to my students. I chose to demonstrate many of these projects using the green Screen by Do Ink app because it is one of the most intuitive and student-friendly apps I have come across for creating green screen content. That being said, with a little effort, each of these projects could be completed with similar apps or tools. Greg Kulowiec coined the phrase *app smashing* some time ago. He writes about the concept on his blog (http://kulowiectech .blogspot.com/) There he describes app smashing as "the process of using multiple apps to create projects or complete tasks." You will see that each of these projects leverages this idea. Dig in and have some fun with these projects. Use them as a starting point with your students. Remix them and then, most important, share them with the world. By doing so, you encourage us all to grow and learn together. I hope that you and your students will have as much fun with these projects as my students and I did!

1. Superhero Project

In this project, students will explore the characteristics of what makes a hero. Each student will develop his or her own superhero persona and will put the resulting character in front of a heroic background image. Finally, each student will create a "museum tag" that will accompany his or her portrait when printed.

Setup

This project is the culmination of a larger study on what it means to be a hero. Students read a variety of picture books, including *Dex: The Heart of a Hero*, written by Caralyn Buehner and illustrated by Mark Buehner; *Superhero ABC*, written and illustrated by Rob McLeod; *Ten Rules of Being a Superhero*, written by Deb Pilutti; and *The Superhero Instruction Manual*, written by Kristy Dempsey and illustrated by Mark Fearing. After reading these or similar books, students generate a list of the characteristics of heroes as well as the causes superheroes tend to strive to protect. Students complete a Superhero Planning Sheet to prepare for their portrait and animation project.

Materials

- Superhero Planning Sheets
- Superhero capes
- Superhero masks

Technology

- Green screen
- Tripod stand
- iPad
- Green Screen by Do Ink app for iPad

Instructions for Students

1. Complete the Superhero Planning Sheet (Figure 6-1).

2. Have a partner photograph you in your best superhero pose in front of the green screen. If you want to "fly," you will need to cover a sturdy table or box with green fabric to "erase" the support. You could also paint the box the same color as your green screen (Figure 6-2).

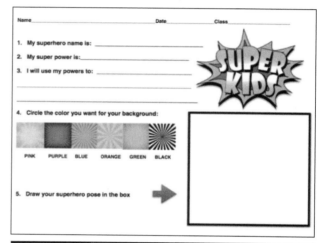

Figure 6-1

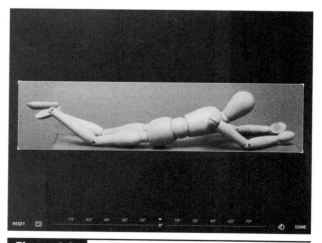

Figure 6-2

Tip: When using poster board as a green screen, it is important to make sure that there are no "hot spots," or areas that are so bright that the green is not visible. This will make it very difficult to remove the background in the green screen software. One way to get around this is to crop the image before importing it into the green Screen by Do Ink app. You can also crop the image within the green screen program, but I usually prefer to do it before importing.

3. Import the color background you've chosen for your superhero into the green Screen by Do Ink app, making it the bottom layer (Figure 6-3).

4. Use the two-finger pinch and zoom maneuver to position the background where you would like it (Figure 6-4).

5. Import the photograph of you against the green screen into the green Screen by Do Ink app (Figure 6-5).

6. Use the two-finger pinch and zoom maneuver to reposition the image the way you would like it.

7. Tap on the green screen layer.

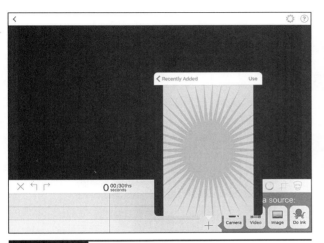

Figure 6-3

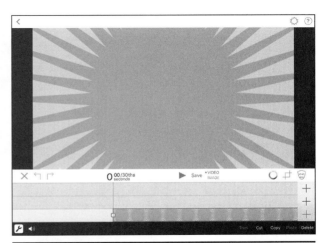

Figure 6-4

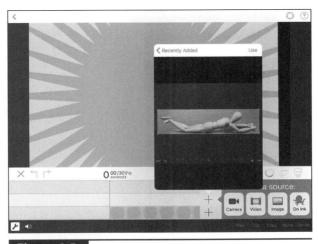

Figure 6-5

8. Use the color wheel to adjust the Chroma key color to match the green of your background. Adjust the sensitivity to dial in the image to your liking (Figure 6-6).

9. When you are pleased with the overall composition, slide the Video/Image tab to Image, and click Save (Figure 6-7).

10. Then save the image to your camera roll.

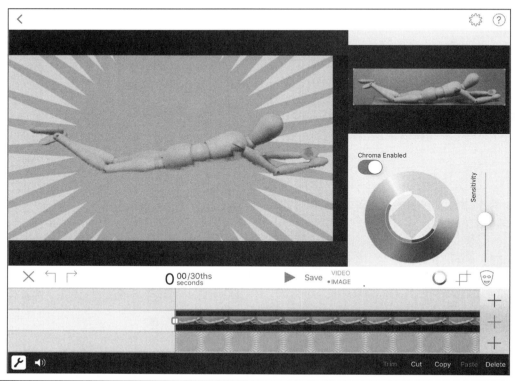

Figure 6-6

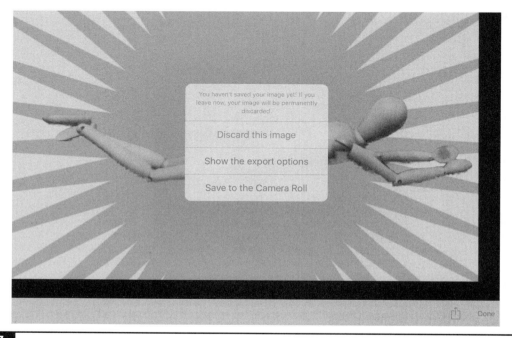

Figure 6-7

Extensions

Tricia Fuglestead inspired this project with her outstanding Superhero Silhouette Poster project. In this project, students use another simple app, Superimpose for iPad, to create a silhouette of their pose and then print their creation. Students could use the iMovie for iPad app and multiple green screen photographs to create "trailers" about their superhero to highlight their superhero's characteristics and mission.

2. *I Spy* Book

In this project, students will create their very own *I Spy* book page and hide themselves in it. A "buddy class" will collaborate with the students by writing the riddle clues for the reader/viewer. The finished project will be shared as a "buddy project," and older and younger buddies will read and explore it together.

Setup

This project is a great way to help students develop literacy and visual acuity. Students will read a wide variety of *I Spy*–type books. My three favorites are *I Spy: A Book of Picture Riddles*, *I Spy Extreme Challenger: A Book of Picture Riddles*, and *I Spy School Days: A Book of Picture Riddles* by authors Jean Marzollo et al. and illustrator Walter Wick. Then students will create several lists of what makes for a fun picture riddle photograph, what makes a good riddle clue, and what objects they want to include in their pictures.

Materials

- A wide assortment of small objects
- Colorful 12- × 18-inch construction paper

Technology

- Green screen
- Tripod stand
- iPad
- Green Screen by Do Ink app for iPad

Instructions for Students

1. Gather 100 small, colorful objects. For younger students, it may be helpful to use tally marks to keep track of the 100 objects.

2. Arrange the objects on a 12- × 18-inch piece of construction paper in a visually interesting style.

3. Use the iPad stand and iPad to photograph your collection from above, making sure to capture all the construction paper and your nametag in the upper-right-hand corner (Figure 6-8).

4. Have a partner photograph you in a fun pose in front of the green screen (Figure 6-9).

Figure 6-8

Figure 6-9

5. Then import the photograph of the objects on the construction paper into the Green Screen by Do Ink app, making it the bottom layer (Figure 6-10).

6. Then import your green screen image, making it the top layer. Use the two-finger pinch and zoom maneuver to position the image where you want it (see Figure 6-11).

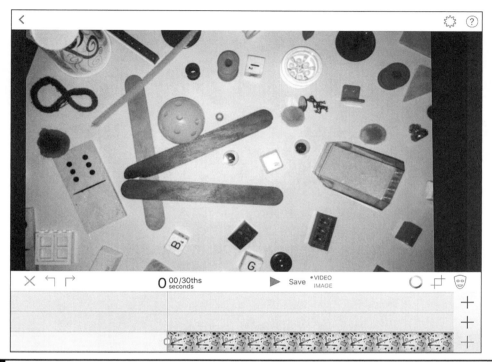

Figure 6-10

Figure 6-11

7. Tap on the green screen layer.

8. Use the color wheel to adjust the Chroma key color to match the green of your background; adjust the sensitivity to dial in the image to your liking (see Figure 6-12).

9. Slide the Video/Image tab to Image, and click Save (see Figure 6-13).

Figure 6-12

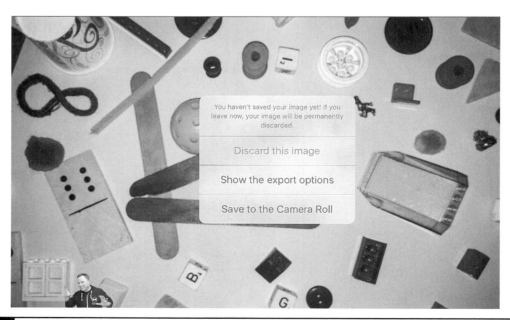

Figure 6-13

10. Then save the image to your camera roll.

11. After printing your picture on legal-sized paper, glue it to the 12- × 18-inch construction paper.

12. After the buddy classes have written, typed, and printed their clues, assemble the photographs and riddles together into a class book.

Extensions

Writing the picture riddles is the most challenging aspect of this project. The 100 objects could be part of the one-hundredth day of school celebration. Another way to make the riddle writing more manageable would be to have the same 100 objects for all students and then have the class generate rhyming words for all the objects.

3. Wonders of the World

In this project, students will explore some of the most majestic places in the world, and they will create travel posters of their destinations, complete with a photograph of each student in front of his or her "wonder."

Setup

This project is a fun way to explore both physical and cultural geography. Students will research the "wonders of the world" and find one to which they would like to travel.

Materials

- Photographs of the many "wonders of the world"
- Poster board
- Markers
- Tape
- Glue

Technology

- Green screen
- Tripod stand

- iPad
- Green Screen by Do Ink app for iPad
- Photo Mapo app for iPad
- Google Earth app for iPad

Instructions for Students

1. Explore the many "wonders of the world" lists, and discuss why some destinations, locations, or sites are on one list but not another.

2. Develop a theory for why something is considered a "wonder," and come up with several characteristics that you think should be considered when naming something a wonder.

3. Come up with a location or site that the group or class would like to explore.

4. Open the Google Earth app.

5. Type in the name of the wonder or the location in the search window on the top right, for example, Machu Picchu (Figure 6-14).

Figure 6-14

Machu Picchu

Historical place in Peru

Machu Picchu is an Incan citadel set high in the Andes Mountains in Peru, above the Urubamba River valley. Built in the 15th century and later abandoned, it's renowned for its sophisticated dry-stone walls that fuse huge blocks without the use of mortar, intriguing buildings that play on astronomical alignments and panoramic views. Its exact former use remains a mystery.

Elevation
7,972'

Area
5,019 m²

Abandoned

Figure 6-15

6. Tap on the red "pin" marked letter "A" to read details about the wonder (Figure 6.15).

7. Drag the orange "person" to the wonder to access "street view" and be able to have a 360-degree view of the location; you'll be prompted to put the figure on the "blue" path where the street-view camera has traveled (Figure 6-16).

8. Click on the highlighted spots to view photographs that other visitors took in this place (Figure 6-17).

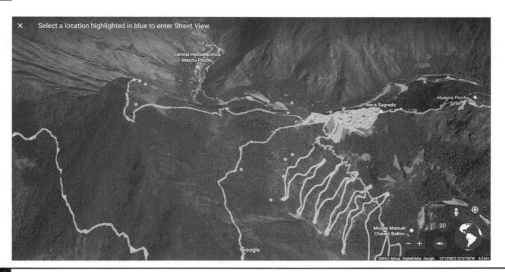

Figure 6-16

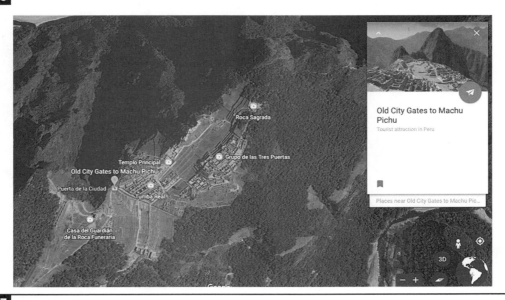

Figure 6-17

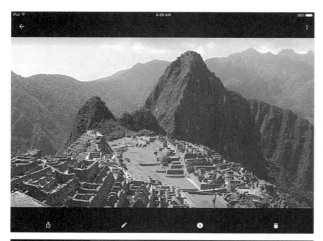

Figure 6-18

Figure 6-20

9. Explore the many options. When you've found one that you would like to use as your backdrop, tap the Share icon on the top right and open the image in Safari (Figure 6-18).

10. Have a partner photograph you in front of the green screen (Figure 6-19).

11. Import your "wonder" photograph into the Green Screen by Do Ink app, making it the bottom layer (Figure 6-20).

12. Import your green screen photograph, making it the top layer (Figure 6-21).

Figure 6-19

Figure 6-21

13. Use the color wheel to adjust the Chroma key color to match the green of your background; adjust the sensitivity to dial in the image to your liking (Figure 6-22).

14. Slide the Video/Image tab to Image, and click Save (Figure 6-23).

15. Then save the image to your camera roll.

16. Next, open Photo Mapo.

Figure 6-22

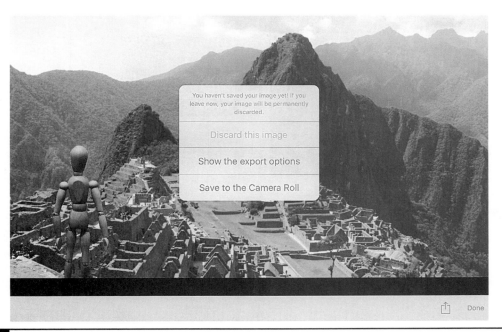

Figure 6-23

Figure 6-24

19. The location will not actually be at the "wonder." You are going to "trick" the program. Enter the photograph location, "Machu Picchu, Cuzco, Peru," and click Search (Figure 6-26).

20. Write a short description of the location (you can only use 180 characters). Machu Picchu is an Incan city built high in the Andes Mountains in Peru. Its elevation is 7,972 feet. It was built in 1450 and abandoned in 1572. Its exact purpose is a mystery (Figure 6-27).

17. Choose a style for your map (Figure 6-24).

18. Select your photograph of you in front of the "wonder" (Figure 6-25).

Figure 6-25

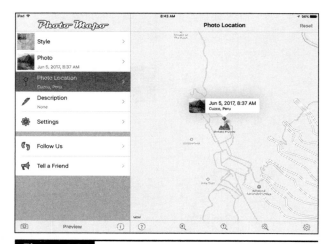

Figure 6-26

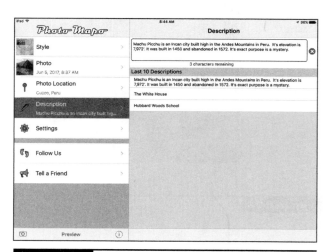

Figure 6-27

21. Tap the Preview icon in the bottom corner for a preview. You can adjust the theme by clicking the left or right arrow button in the top-right corner (Figure 6-28).

22. When you are happy with your composition, click the Share icon at the bottom, and save the image to the camera roll.

23. Print your Photo Mapo design and add additional images, details, or maps to complete your project (Figure 6-29).

Extensions

Students could use the Photo Mapo project as a background for a video report, using it as the bottom layer in the Green Screen by Do Ink app. Students could also compile the "wonders" into an e-book using the Book Creator for iPad app. They could add a link to the video using a QR code as well.

Figure 6-28

Figure 6-29

4. You're on the Cover of a Magazine!

Being on the cover of a magazine is something that very few of us have the chance to experience. In this project, students not only will appear on the cover of a magazine, but they will also develop all the details about their magazine, including headlines, pricing, and other thoughtful features. As is the case with many projects, there are a number of ways to do this activity. These include complex programs such as Photoshop and versions of templates made in word-processing and/or presentation software such as Keynote. For the sake of ease, I've chosen to use Big Huge Labs' Magazine Cover template (https://bighugelabs.com/magazine.php). While there are a multitude of options, this is one of the simplest ways to do this project.

Setup

There are magazines that cater to just about any interest group, hobby, or activity. Depending on the age group of your students, you might want to presearch magazine covers to screen out inappropriate content. You might copy some that cater to popular topics or hobbies of your students such as *Hockey*, *Equestrian*, *Boy's Life*, *Girl's Life*, *National Geographic*, *Sports Illustrated*, *TIME*, or *People*.

Materials

- Photocopies or physical copies of magazine covers related to your students' interests
- Props and costumes for cover photographs

Technology

- Green screen
- Tripod stand
- iPad
- Green Screen by Do Ink app for iPad

Instructions for Students

1. Study the magazine covers and begin to analyze them for details:

   a. What do you notice about the photographs?

   b. What do you notice about the magazine titles?

   c. What are the common elements on the magazine covers?

2. After making a list of the characteristics and common elements, imagine that you could be on the cover of any magazine. Which magazine would it be?

3. Complete the Magazine Cover Planning Sheet (Figure 6-30).

Magazine Cover Planning Sheet

What is the title of your magazine?

_____

What is the magazine's tagline? For example: The Most Important News for the Most Important People in the World!

_____

Publication Date:

_____

Price:

_____

Lines 1-10 are optional. You can make up headlines that might have something to do with you, or to just make the cover appear more realistic to your audience.

Lines 11-13 should give a little more information about why it is you are on the cover.

_____

_____

Lines 14 and 15 should give additional details about you and your story.

_____

_____

Lines 16 and 17 appear at the bottom of the cover and are usually in the colored portion. They are optional as well.

Figure 6-30

4. Before you can create your magazine, you have to capture the cover photograph. Have a partner photograph you in front of the green screen; you might want to bring in props to help you "sell" your cover photo. For example, if you want to be on *Runner Magazine*, you might want to wear your running shoes and gear (Figure 6-31).

5. Search for a background image that will enhance your cover photo. It can be a simple colored background or a photo of where you might want to "put" yourself for your cover photo (Figure 6-32).

Figure 6-33

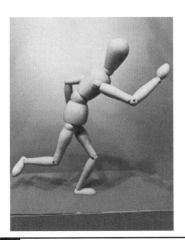

Figure 6-31

6. Import your background image into the Green Screen by Do Ink app, making it the bottom layer (Figure 6-33).

7. Import your green screen photograph, making it the top layer.

8. Use the color wheel to adjust the Chroma key color to match the green of your background; adjust the sensitivity to dial in the image to your liking (Figure 6-34).

Figure 6-32

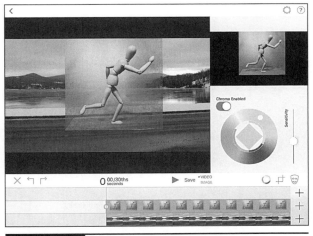

Figure 6-34

9. Slide the Video/Image tab to Image, and click Save (Figure 6-35).

10. Then save the image to your camera roll (Figure 6-36).

11. Open Big Huge Labs Magazine Cover template (https://bighugelabs.com/magazine .php).

12. Use your Magazine Cover Planning Sheet to help you fill in the details of your magazine cover.

Figure 6-37

13. Upload your photograph (Figure 6-37).

14. You need to decide how you want to crop your image. This is not easily perfected, so give one of the options a try. You can always choose another if you are unhappy with the result. You will receive a preview in the small window at the bottom right side (Figure 6-38).

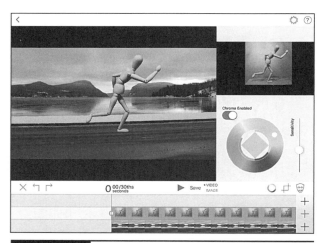

Figure 6-35

Figure 6-36

Figure 6-38

15. Next, choose a layout; the ones shown are merely suggestions and can be tweaked as well.

16. Depending on the layout, some lines may be more or less important. Check the preview to see which lines are the most prominent.

You need to decide which font you would like your magazine to use (this is one weakness of this template; you can only use one font) (Figure 6-39).

17. For this template, the most important lines are 7, 8, and 9 (Figures 6-40 and 6-41).

Figure 6-39

All of the following fields are optional.

Figure 6-40

All of the following fields are optional.

Magazine title

Runner's World

Bold ☑ Shadow ☑ Color ☐ #ffffff

Tagline

We Keep You in Front of the Pack!

Bold ☑ Shadow ☐ Color ☐ #ffffff

Publication date

June 4, 2017

Bold ☑ Shadow ☐ Color ☐ #ffffff

Price

$6.99

Bold ☑ Shadow ☐ Color ☐ #ffffff

Figure 6-41

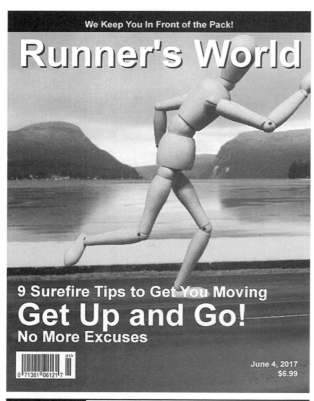

Figure 6-42

18. When you've added your details, click Create (Figure 6-42).

19. What's wonderful about this site is that by clicking the back arrow, you can make any changes you like and then "remix" the project without having to reenter all the information.

20. When you are happy with the composition, download the cover.

21. Done. You are famous!

Extensions

Each year, *Time* magazine chooses a "Person of the Year." Students could think about how they might influence the world for good. They could imagine themselves in 20 years being featured as *Time*'s Person of the Year. Their headlines and additional story details could be written to reflect their work in making the world a better place. Students could also edit their photographs to put them "back in time" and put them on the cover of a historical magazine at a specific historical moment.

5. Eye to Eye with the BFG, or the "Big Friendly Giant"

In this project, students will put themselves in the hand of the beloved giant from Roald Dahl's *The BFG*.

Setup

The BFG by Roald Dahl is one of the most popular books I've ever shared with students. The film adaptation has caused even more students to fall in love with Roald Dahl. After reading the book, students will all want to stand in the giant's hand and look him in the eye.

Materials

- Image of the giant from *The BFG*
- Milk crate or similar sturdy box

Technology

- Green screen
- Green cloth for the ground
- Tripod stand
- iPad
- Green Screen by Do Ink app for iPad

Instructions for Students

1. Create your own illustration of the giant (Figure 6-43).
2. Photograph your drawing of the BFG.
3. Have a partner photograph you in front of the green screen; be sure that you pose as if you are looking directly into the eyes of the

Figure 6-43

Figure 6-44

giant. Also, make sure that you stand on the milk crate covered in green fabric so that the illusion of the giant's hand is maintained (Figure 6-44).

4. Import your background image (illustration of the giant) into the Green Screen by Do Ink app, making it the bottom layer (Figure 6-45).

5. Import your green screen photograph, making it the top layer (Figure 6-46).

Figure 6-47

Figure 6-45

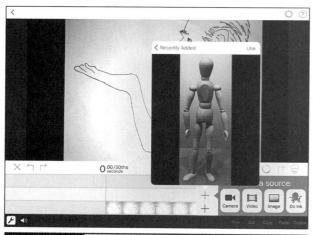

Figure 6-46

6. Tap the top layer, use the two-finger pinch to shrink the size of your photograph, and move it to the "hand" of the giant (Figure 6-47).

7. Use the color wheel to adjust the Chroma key color to match the green of your background; adjust the sensitivity to dial in the image to your liking (Figure 6-48).

Figure 6-48

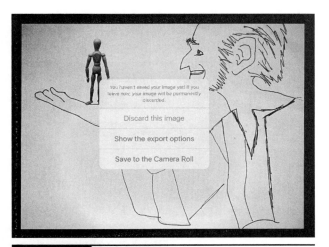

Figure 6-49

Extensions

Students could use video instead of stills. Students could act out their favorite lines between the child and the BFG. Students could also "be" the giant by combining two video layers in the Green Screen by Do Ink app. Just use the same two-finger pinch and zoom to make the child seem small in the hands of the BFG. This is a fun technique for just about any "monster" or giant. Once students master it, they will want to try all sorts of permutations.

8. Slide the Video/Image tab to Image, and click Save (Figure 6-49).

9. Save the image to your camera roll.

10. Print your photograph of the BFG and you eye to eye!

6. Snow Globe Poetry Project

This project comes from educator Jessica Goodrow from Conroe, TX. In it, students will use the green screen to transport themselves inside a snow globe. They will create their own wintry snow globe and a wintry poem to accompany their image.

Setup

Wintry snow globes can transport us instantly to a magical snowy day with just a shake. Students will love dressing up in their winter gear and "trapping" themselves inside a snow globe. They'll also write a poem about winter to accompany their photograph.

Materials

- Snow globes from a variety of locations
- Winter clothing
- Snow Globe Poem Sheet
- Snow globe image (https://pixabay.com/en/crystal-ball-glass-globe-glass-ball-32381/)

Technology

- Green screen
- Tripod stand
- iPad
- Green Screen by Do Ink app for iPad

Instructions for Students

1. Explore a variety of snow globes either real or virtually via the Internet.

2. Dress in your winter gear, and have a partner photograph you in front of the green screen.

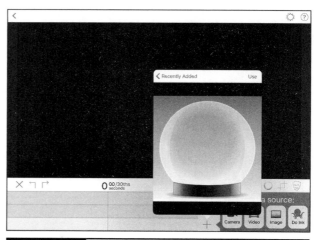
Figure 6-50

3. Import your background image (snow globe image) into the Green Screen by Do Ink app, making it the bottom layer (Figure 6-50).

4. Import your green screen photograph, making it the top layer (Figure 6-51).

5. Tap the top layer, use the two-finger pinch to shrink the size of your photograph, and move it inside the snow globe.

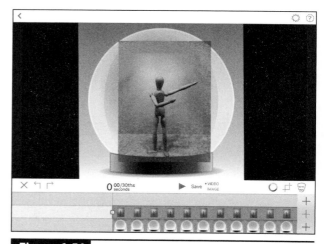
Figure 6-51

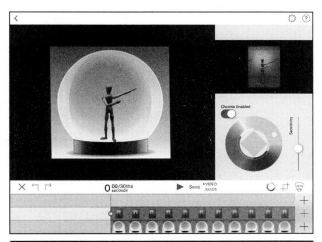

Figure 6-52

6. Use the color wheel to adjust the Chroma key color to match the green of your background; adjust the sensitivity to dial in the image to your liking (Figure 6-52).

7. Slide the Video/Image tab to Image, and click Save (Figure 6-53).

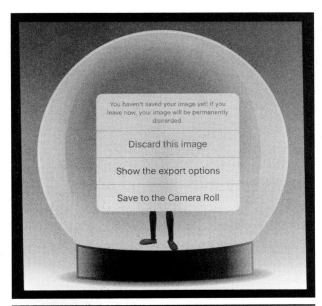

Figure 6-53

8. Save the image to your camera roll (Figure 6-54).

9. Print your photo.

10. Complete the five senses poem to describe what it is like inside the snow globe (Figure 6-55).

11. Display the photo of you inside the snow globe along with your sensory poem.

Figure 6-54

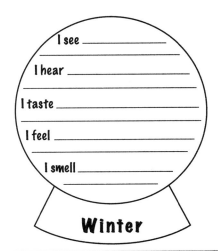

Figure 6-55

Extensions

Students can make a physical snow globe with this do-it-yourself (DIY) project: www.marthastewart.com/1099137/diy-bell-jar -snow-globes. Rather than simply using a static image, students could use the Animation by Do Ink app to animate the snowflakes inside their snow globes and create a video of themselves inside the snow globe too. Students can learn more about the family that invented the snow globe at www.bbc.com/news/business-25298507.

7. X-Ray Machine

This project came from fourth grade teacher Heidi MacGregor from Russel Street School in Littleton, MA. In this project, after studying the human skeletal system, students create lifelike "x-rays" of their bodies (Figure 6-56).

Figure 6-56

Setup

This project will help students to identify the various parts of the skeletal system in order to create a realistic looking x-ray.

Materials

- Images of x-rays of the human body (Here is a link to get you started: http://maxpixel.freegreatpicture.com/static/photo/1x/Human-Skeleton-Body-Medical-Anatomy-Bones-Skeleton-1813086.jpg.)

- Whiteboard with a square of green construction paper taped to it

Technology

- iPad
- Green Screen by Do Ink app for iPad

Instructions for Students

1. Identify which part of your body you would like to "x-ray."

2. Have a partner photograph you with the framed Chroma key green foam sheet over the part of your body you are going to "x-ray" (Figure 6-57).

3. Download the x-ray image (Figure 6-58).

Figure 6-57

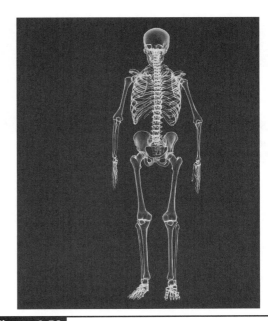

Figure 6-58

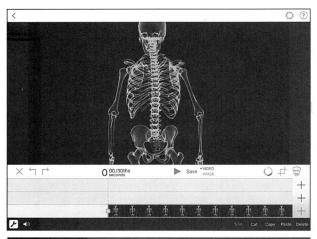

Figure 6-59

6. Adjust both images using the two-finger pinch and zoom.

7. Use the color wheel to adjust the Chroma key color to match the green of your background; adjust the sensitivity to dial in the image to your liking.

8. Slide the Video/Image tab to Image, and click Save.

9. Save the image to your camera roll (Figure 6-61).

10. Print your photograph (Figure 6-62).

4. Import the x-ray image into your green screen, making it the bottom layer (Figure 6-59).

5. Import the photograph of you with the picture frame with green foam in it, making it the middle layer (Figure 6-60).

Figure 6-61

Figure 6-60

Figure 6-62

Extensions

Students could do this for any of the body's other systems as well. Imagine looking "through" the skin to see the heart, lungs, or brain. Why stop at still images? Use a video of a beating heart or lungs that are expanding and contracting. John Kline, director at WeVideo for schools, even has a "portable green screen T-shirt" that he uses to demonstrate how easily WeVideo can be used to Chroma key on the go. Link to tweet: https://twitter.com/EDUcre8ive/status/880548610793807872.

8. Monument Project

In this project, students will put their faces on famous world monuments. This would be a terrific way to deliver oral reports that is also fun (Figure 6-63).

Setup

After studying several human-made monuments such as the Sphinx, Crazy Horse, Mount Rushmore, the Lincoln Memorial, etc., students will put their faces onto monuments of their choice. Students will present their research via video "from" the monument.

Materials

- Photographs of various monuments
- Research on monuments

Technology

- Tripod stand
- iPad
- Green Screen by Do Ink app for iPad

Figure 6-63

Instructions for Students

1. Choose your monument, and do your research.

2. Find an image of your monument.

3. Study the orientation of the photograph to plan how you need to position yourself while recording your oral presentation.

4. Record your oral presentation, being very careful not to move your head.

5. Import the image into Green Screen by Do Ink app, and *rather than making it the bottom layer, make it the middle layer* (Figure 6-64).

6. Tap on the middle layer to make it active.

Figure 6-65

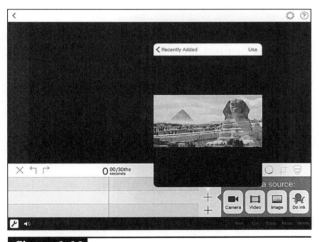

Figure 6-64

7. Click on the Small Mask icon. This opens up a new window with five tools. Select the eraser tool, and adjust the opacity of the image (Figure 6-65).

8. Use the eraser tool to remove the "face" on the monument. You may need to adjust the opacity to make the illusion appear more realistic.

9. Import the video/photo that you want to put on the monument, making it the *bottom layer* (Figure 6-66).

Figure 6-66

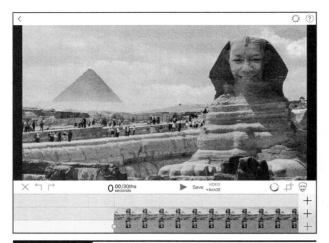

Figure 6-67

Figure 6-68

10. Use the two-finger pinch to shrink the size of your face to that of the face on the monument (Figure 6-67).

11. You may need to adjust the length of the video of the monument to match the video of your oral presentation. Do this by tapping on the layer and extending the length on the timeline.

12. Slide the Video/Image tab to Video, and click Save (Figure. 6-68).

13. Save the image to your camera roll.

Extensions

Students could do the same thing with famous paintings or photographs. Imagine your students using the same technique to create "holes" for each of the crew members and George Washington in the famous painting *Washington Crossing the Delaware* by Emanuel Leutze.

9. Immigration Family Pictures at Ellis Island and Map Project: Postcard App

In this project, students will create a postcard of their immigrant family at Ellis Island on the day they entered the United States (Figures 6-69 and 6-70).

Figure 6-69

Ellis Island
We made it! We are safely here!

Dear family,

After two long weeks in steerage, we arrived in America. Louisa was put in quarintine for her cough, but after a few days she got better. Now we are excited to move in with uncle Ralph and start saving to move you all here!
With love,
Salvatore

Hubbard Woods School
1110 Chatfield Road
Winnetka, IL 60093

Figure 6-70

Setup

This project was part of the culmination of an interdisciplinary exploration of American immigration. Students participated in a day-long simulation in which they experienced many of the same challenges immigrants faced when they came through Ellis Island. They came dressed in simple clothing, and each "family" created a basic suitcase in which they had to place very few precious belongings. This project helped cement the experiences the students had and served as an excellent way for students to consolidate their learning.

Materials

- Images of Ellis Island (A good source is *At Ellis Island: A History in Many Voices* written by Louise Peacock and illustrated by Walter Lyon Krudop.)

Technology

- Green screen
- Tripod stand
- iPad
- Green Screen by Do Ink app for iPad
- PostCard Creator app or online

Instructions for Students

1. After reading and discussing *At Ellis Island: A History in Many Voices*, form "family groups" and create a postcard to send "back" to the family that was left when your group immigrated to the United States.

2. Pose as a family in front of the green screen background (Figure 6-71).

Figure 6-71

Figure 6-72

3. As a family, choose a "background" photograph you would like for your postcard.

4. Import an Ellis Island background image, making it the bottom layer (Figure 6-72).

Figure 6-73

Figure 6-74

5. Import your "family portrait" taken against the green screen, making it the top layer (Figure 6-73).

6. Use the color wheel to adjust the Chroma key color to match the green of your background; adjust the sensitivity to dial in the image to your liking.

7. Slide the Video/Image tab to Image, and click Save (Figure 6-74).

8. Save the image to your camera roll.

9. Convert the image to either black and white or sepia to further enhance the historic nature of the photograph (Figure 6-75).

Figure 6-75

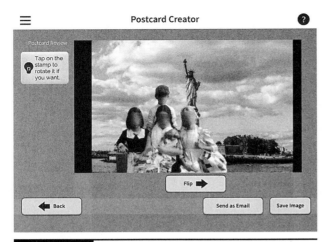

Figure 6-76

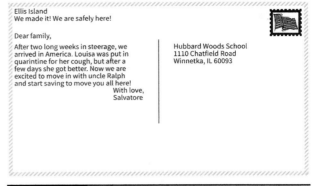

Figure 6-77

10. Use online tools or the PostCard Creator app to create and print your postcard (Figures 6-76 and 6-77).

11. Glue your "family photograph" on cardstock, and glue the text and image together.

Extensions

Students could augment this project by using the Photo Mapo app to show where their families immigrated from and where they arrived in the United States. Students could also find historical photographs of the city where they sailed from on their way to Ellis Island and could create several memories from various parts of their journey.

10. Word Cloud Project

In this project, students create a "word cloud" and then put themselves in front of it as a gift for a loved one (Figure 6-78).

Setup

This project is a fun way to personalize a unique gift for loved ones. The Word Cloud app uses the frequency of words to determine their size. Students can use words they want to appear larger multiple times to enhance this effect. The final result is a personalized word cloud.

Materials

- Brainstorming sheet
- Word bank

Technology

- Green screen
- Tripod stand
- iPad
- Green Screen by Do Ink app for iPad
- ABCya online or the ABCya app

Figure 6-78

Instructions for Students

1. Brainstorm descriptive words about the person for whom you want to make a gift (Figures 6-79 and 6-80).

affectionate,
amusing,
beautiful,
bigheart,
caring,
charming,
comedic,
compassionate,
considerate,
creative,
dependable,
empathetic,
energetic,
enterprising,
entertaining,
forgiving,
fun,
funny,
generous,
gentle,
giving,
graceful,
great cook,
hardworking,
helpful,
honest,
humorous,
imaginative,
inspiring,

intelligent,
interesting,
inventive,
joyful,
kind,
lighthearted,
loving,
loyal,
nice,
nurturing,
outspoken,
patient,
playful,
positive,
practical,
proper,
rolemodel,
special,
strong,
sweet,
sympathetic,
thoughtful,
trustworthy,
uncompromising,
understanding,
unique,
unselfish,
wise,
wonderful.

Figure 6-79

Name_____Date_____

Who are you going making your word cloud portrait for?_____
Think of 20 descriptive words to describe this person:

Figure 6-80

2. Use either the iPad app or the online version to type in each of the descriptive words (Figure 6-81).

motherly
playful
forgiving
kind
Caring
awesome
helpful
giving
self-sacrificing
Mom
Mom
Mom

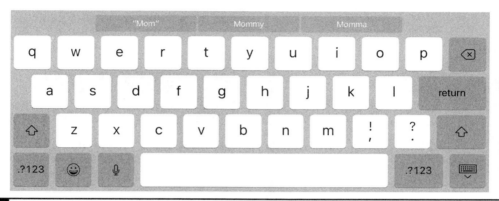

Figure 6-81

3. Enter the name of the person for whom you are making the gift more than any other word. This will ensure that the recipient's name is the largest of all the words.

4. After entering all the words, tweak the settings for color, font, and orientation (Figure 6-82).

5. When you are pleased with all the variables, save the design to your camera roll (Figure 6-83).

6. Next, have a partner photograph you in front of the green screen.

Figure 6-82

Figure 6-83

7. Import your word cloud image into the Green Screen by Do Ink app, and resize it by selecting the layer and pinching and zooming to fit the entire image in the viewer. Make this layer the bottom layer (Figure 6-84).

8. Then import your green screen photograph into Green Screen, placing it on top of the word cloud layer (Figure 6-85).

9. Pinch and zoom the image to place yourself to the right or left of the word cloud.

10. Use the color wheel to adjust the Chroma key color to match the green of your background; adjust the sensitivity to dial in the image to your liking (Figure 6-86).

11. Slide the video/Image tab to Image, and click Save (Figure 6-87).

12. Save the image to your camera roll.

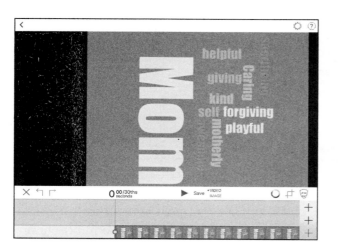

Figure 6-84

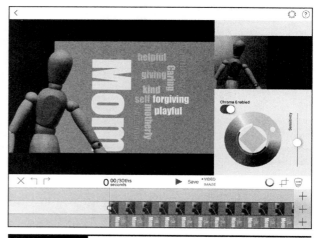

Figure 6-86

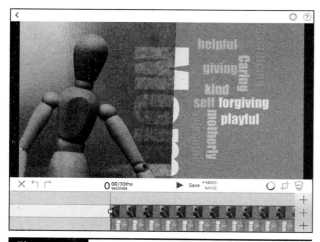

Figure 6-85

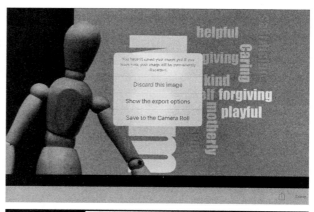

Figure 6-87

Extensions

Tagxedo is another word cloud generator that allows you place your word cloud inside a particular shape, say, a star, a circle, or even a silhouette from an image. It's a bit more complex than the ABCya app, but the results are quite impressive. Students could also insert a video of themselves talking about why their loved one is so important to them.

11. "Live" Front Page of a Newspaper: The Titanic Sinks!

In this project, students will create an animated "live" version of a famous newspaper front page, much like those found in the Harry Potter films (Figure 6-88).

Setup

Students will choose a famous historical event and create an animated version of the front page of the newspaper from the day the event occurred. This project uses one of the powerful editing features of the Green Screen by Do Ink app, the masking tool. Once students see how easily they can mask out certain areas of a background, they will find all sorts of new uses for the app.

Materials

- Real or virtual copy of the newspaper for the day the students wish to bring to life
- Props to fit the event
- Background image (A good source is "TITANIC SINKS, 1500 DIE—Carpathia Picks Up 675 Out of 2200—Races for New York—Survivors Mostly Women and Children," *Boston Daily Globe*, April 16, 1912, p. 1: print.)

Technology

- Green screen
- Tripod stand

Figure 6-88

- iPad
- Green Screen by Do Ink app for iPad

Instructions for Students

1. Search for important events in world history as seen on the front pages of newspapers (Figure 6-89).

2. Set up the scene you want to put on the cover of your newspaper, and film it in front of the green screen.

3. Import that footage or photograph into the Green Screen by Do Ink app, making it the bottom layer (Figure 6-90).

Figure 6-89

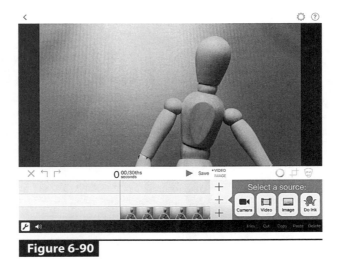

Figure 6-90

4. Import the newspaper image into the green screen, making it the top layer (Figure 6-91).

5. Click on the mask tool (Figure 6-92).

6. Use the eraser tool to remove the part of the scene you are going to recreate (Figure 6-93).

7. You can adjust the sensitivity, size, and opacity of the eraser as well (Figure 6-94). This essentially creates a "hole" that you can see through to the layer that is on the bottom (Figure 6-95).

Figure 6-93

Figure 6-91

Figure 6-94

Figure 6-92

Figure 6-95

8. Resize the bottom layer to fit through the hole.

9. When you are pleased with the composition, slide the Video/Image tab to Video, and click Save (Figure 6-96).

10. Save the image to your camera roll.

Figure 6-96

Extensions

Students could create their own newspaper front page and then animate the headlines. They could animate the headlines with stop motion and then create looping illustrations with an online GIF maker (https://giphy.com/create/gifmaker).

12. TouchCast PSA Project

This project comes to us from Laura Gardner, teacher/librarian at Dartmouth Middle School in Dartmouth, MA, and TouchCast STAR Ambassador. In this project, students will create a 30-second to 1-minute public service announcement (PSA) using TouchCast. For example, students could create PSAs about social issues or digital citizenship topics for their peers, parents, and community. Creating a PSA requires research, collaboration, and speaking skills. TouchCast is an excellent tool for this project; the tool includes a teleprompter, a whiteboard, a green screen option, and virtual applications (vApps) that can make your video interactive and engaging. In addition, TouchCast is free for educators and can be exported to YouTube, the camera roll, and social media, as well as your own TouchCast channel. This app is also a great option to use for a morning news show, flipping the classroom, and any other curricular project.

Setup

Much of the work for this project is done before filming. Before starting the project, introduce the idea of PSAs by watching ads from the Ad Council (www.adcouncil.org/Our-Campaigns) and analyzing them (https://goo.gl/fbVCvh).

Materials

- Creativity
- Storyboard

Technology

- Green screen
- Tripod stand
- iPad
- TouchCast app for iPad

- Computer
- Google Docs
- Google Drive
- PVLEGS rubric (http://pvlegs.com/effectiverubrics/pvlegsrubric/)

Instructions for Students

1. Brainstorm! After viewing other PSAs, it's time to make your own! Typically, groups of three to four students must brainstorm the problem they want their PSA to address. Questions you might ask yourselves include

   a. What is the problem?

   b. Why is it a problem?

   c. How does it affect individuals and others?

   d. What sources might we use to research this further?

   e. What are potential solutions?

2. Then you should select at least one ad technique and at least two effects for your PSA. For example:

Ad Technique (must pick at least one)	Effects You Could Use (must pick at least two)
Appeal by association	Fade in/out
Humor	Music (10 percent or 30 seconds of a copyrighted song is okay as long as you cite it!)
Fear	
Loaded language	
Emotional appeal	Silence
Repetition	Time lapse (requires app smashing)
Catchy slogan	
Cute factor	Sound effects
Bandwagon	Echo
Story	Change of scenery
	Zoom in/out
	Camera angles
	Voice-over
	Words on screen
	Filters

Storyboard! Now it's time to use a storyboard to flesh out your ideas. Scholastic has a great storyboard for PSAs (www.scholastic.com/drive2life/pdf/ NRSF_612_Graphic%20Organizer.pdf). Think outside the box. Do you want your PSA to be fully acted out, or will portions include images or words on the screen with a voice-over?

3. Script and choose images. If you will be speaking a lot in your PSA, then you should write a script in Google Docs. In addition, you will need to find images for your background and any vApp images you want to pop up on the screen. Use copyright-free images for your background. Choose "images labeled for reuse" in Google (on a computer, click Tools and then Usage Rights; these options are not available on an iPad), or better yet, use a database such as Britannica ImageQuest, which has thousands of images that have been cleared of copyright. You should put citations and links to images in the Google doc.

4. Set up TouchCast. When you are finished with your script, it is time to set up TouchCast. Open Docs on your iPad, and click Select All. Copy the entire script (Figure 6-97).

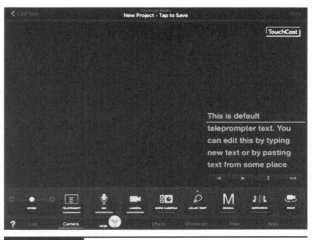

Figure 6-98

5. Then open TouchCast, and click on Camera at the bottom and then Teleprompter to paste the script in (Figure 6-98).

6. Click on the blue Edit button in the teleprompter, press down in the white box, and select Paste (Figure 6-99).

7. Once the script is in the teleprompter, you can click on the "T" in the teleprompter to change the size and color of the teleprompter text. The teleprompter will scroll automatically when recording. If it scrolls too fast or too slow, just adjust the scrubber on the far left of the screen next to the word "Teleprompter."

Figure 6-97

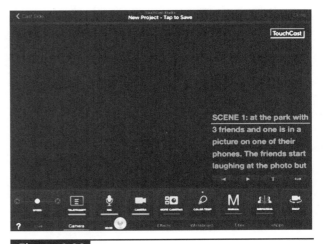

Figure 6-99

8. Now it's time to set up the green screen. Open your Google doc again on the iPad, click the link for each background image, and save those images to the camera roll.

9. In the TouchCast app, click Effects at the bottom and then Green Screen (Figure 6-100).

10. Then click the Camera button on the right side of the screen (iPad 4 and iPad Air models currently allow video backgrounds as well) (Figure 6-101).

11. On the next screen, click on the word "Albums" to find the pictures from the camera roll (TC backgrounds are also available) (Figure 6-102).

12. Once you click on the image from the camera roll and hold your iPad up to a green screen, your background will appear (Figure 6-103).

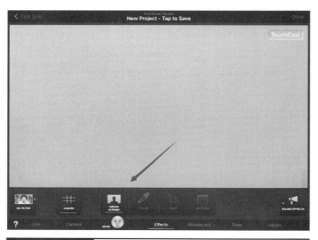

Figure 6-100

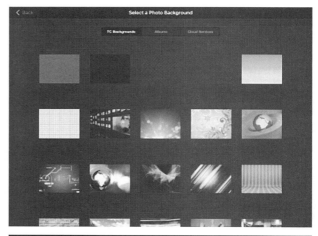

Figure 6-102

Figure 6-101

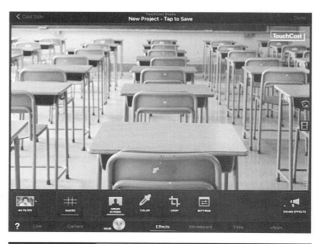

Figure 6-103

13. If you have vApp, you might want text to appear in your video, so you should set those up as well. For example, in some digital citizenship PSAs, students have included screenshots of text messages going back and forth to show cyber bullying. To do this, students screenshot their text messages, added the images to Google Drive on their phones, and then opened those images in TouchCast as vApp photographs. You can also use vApps such as polls, YouTube videos, questions, a web page, and much more (Figure 6-104).

14. You can add vApps when recording as a small pop-up, full screen, or full screen with a small window for your camera. For older iPads, vApps must be added while filming, but for iPad 4 and iPad Air, vApps can be burned into the video during the editing

Figure 6-105

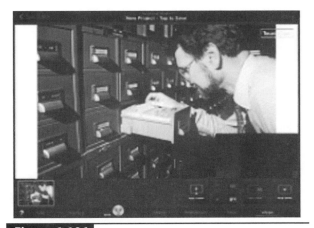

Figure 6-106

process (Figure 6-105). (Note the camera option with full screen in Figure 6-106.)

15. Film your PSA! You should first review the PVLEGS rubric to make sure that you understand the expectations for speaking skills during this project (http://pvlegs.com/effectiverubrics/pvlegsrubric/). You should also make sure that you are effectively using at least one ad technique and some special effects.

16. You do not need long in front of a green screen for this project provided that you have set everything up in advance and have practiced. Editing options are good enough that you should film straight through, repeating sections where you make mistakes.

Figure 6-104

17. You should film "selfie style" so that you can see how you look on camera and make sure that you are referencing your green screen background correctly. Click Camera and then Swap on the far right to swap the camera if needed. Position the iPad on the tripod/iPad stand so that it is stable.

18. To film press the red Record button. To pause, press it again.

19. You can also use the whiteboard to film. You can set up the whiteboard or chalkboard and then type or write on the screen. You need to record the whiteboard for as long as you wish the words to be on the screen. If you want the words to be accompanied by silence, be sure to click Camera and turn off the microphone. (Make sure that students use this feature to cite their image sources and any other sources as a final slide) (Figure 6-107).

Figure 6-107

20. To view your clips, press down in the top-right corner. I do not recommend that you watch your clips until the editing process, however. If you wish to include vApps as you film, you should click on vApps at the bottom before pressing the Record button, and you should click on those vApps one by one at whatever point in filming you want them to show up. Be sure to choose their orientation before filming (Figure 6-108).

21. If you click Done and want to film more, you can click on Add Clip in the bottom-left corner and then Record in Studio with Camera (Figure 6-109).

Figure 6-108

Figure 6-109

22. Editing! Once all the clips are filmed, it's time to edit. Click Done in top-right corner, and watch each clip with an eye to trim. Click on the clip, and then click Trim at the bottom.

23. You can use the scissors to cut a clip into pieces, and then you can drag the blue ends of each clip to trim off sections (Figure 6-110).

24. This is the point at which those of you with newer iPads can add vApps to your clips as well. To do this, simply click on the clip you wish to add a vApp to, and click Add vApps at the bottom of the screen (Figure 6-111).

25. Be sure to set vApps up and then hit Record on the clip, watch your already-recorded scene, and add the vApps one by one in the correct spots.

26. Export! Once your clip is finished, it's time to export your TouchCast. Click Export TouchCast on the editing screen. If you wish to only save the TouchCast to the camera roll, which is a great option for younger students, do not sign in and just click Save to Camera Roll. Otherwise, sign in and choose the options that work best for you (Figure 6-112).

27. Reflect and assess. You should reflect on this assignment. What worked well? What was difficult? What would you do differently next time?

28. Teachers can assess for content using a teacher-created rubric, as well as speaking skills using the PVLEGS rubric.

Figure 6-110

Figure 6-111

Figure 6-112

Works Cited

Classroom. Photograph. Britannica ImageQuest, *Encyclopedia Britannica*, May 25, 2016. Available at quest.eb.com/search/105_1401480/1/105_1401480/cite. Accessed June 9, 2017.

Library. Photography. Britannica ImageQuest, *Encyclopedia Britannica*, May 25, 2016. Available at quest.eb.com/search/119_1810117/1/119_1810117/cite. Accessed June 9, 2017.

13. Stop-Motion Animation Project on Lego Green Screen

In this project, students will create a stop-motion film and backdrop layer that augments the project.

Setup

Stop-motion animation is a fascinating pursuit. Show students some examples of stop-motion animation such as *Wallace and Gromit* or the myriad of Lego stop-motion videos, and they will be incredibly excited to make their own. A YouTube video provides a terrific explanation (www.youtube.com/watch?v=wVjMFU11hVA). In films, there are typically 24 frames per second. Students very quickly realize the importance of tiny movements and many exposures to create smooth, "realistic" animation.

Materials

- Something to animate (Lego minifigures are a great way to start)
- Green or blue Lego base plates
- Bookends to hold the vertical base plates in place
- Tape
- Legos to create a set or vehicles
- Storyboard sheet

Technology

- Boinx iStopMotion app for iPad
- Tripod stand
- iPad
- Green Screen by Do Ink app for iPad

Instructions for Students

1. Use the storyboard sheet to roughly plan out the action, setting, and perspective of each scene (Figure 6-113).

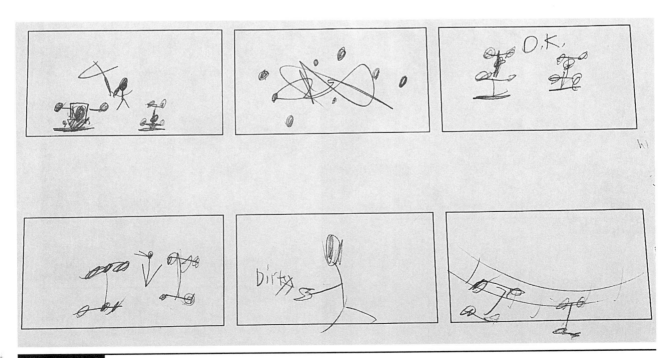

Figure 6-113

2. Select images for each scene. It is important that you do this to make sure that the animation lines up with any details in your scene.

3. Set up the background. I used bookends and tape to anchor both the green and gray base plates (Figure 6-114).

4. Set up the camera on the tripod stand. It is essential that the iPad remain as still as possible to get the best results.

   Tip: Use tape to mark the edges of what is visible from the iPad (Figure 6-115).

5. Photograph the animated object, and then make tiny adjustments and photograph again. This is repeated over and over to achieve the illusion that the animated object is moving on its own (Figure 6-116).

6. Play back the frames to see if the object is moving as you would like it to move.

Figure 6-115

Figure 6-114

Figure 6-116

Figure 6-117

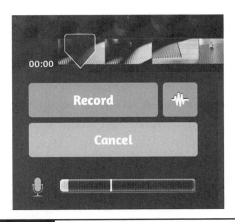

Figure 6-119

7. You can adjust the number of frames per second by clicking on the Gear icon in iStopMotion. Adjusting the number of frames slows down the animation. You can also select Play at Half Speed to slow down the animation to make the movement more fluid. (Figure 6-117).

8. It is best to add sound effects before exporting the film to the camera roll. To access the audio features in iStopMotion, tap on the music note at the bottom-left corner of the screen (Figure 6-118).

9. To record sound effects or dialogue, tap the Record button (Figure 6-119).

10. Place the play head at the spot where you wish to add audio, and click Record. If you would like to import audio, you have several options (Figure 6-120).

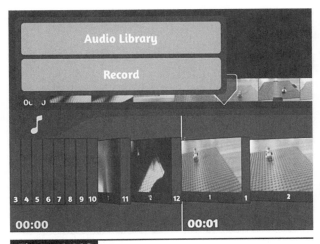

Figure 6-118

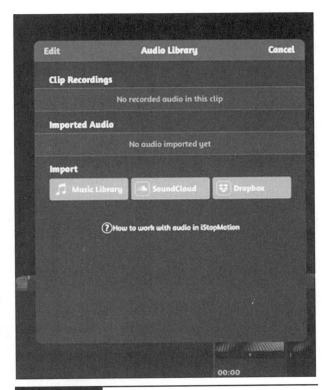

Figure 6-120

11. When you are satisfied with both your animation and your sound effects, save the film as a movie file to the camera roll (Figure 6-121).

12. Open Green Screen by Do Ink app and import the photograph for the first scene in your film, making it the bottom layer (Figure 6-122).

13. Import the video you created with the Boinx iStopMotion into the green screen, making it the top layer; pinch and zoom to size.

14. Use the color wheel to adjust the Chroma key color to match the green of your background; adjust the sensitivity to dial in the image to your liking.

15. When you are satisfied with the timing, slide the Image/Video slider to Video, and Save to save it to the camera roll.

    Tip: Because this video has multiple scenes, tap the background layer and grab the red "handle" to adjust the length that the background image is displayed. For each new scene, import a new background image and adjust its length to fit the timing you would like (Figure 6-123).

16. When you are happy with the timing for each of the scenes, slide the Video/Image tab to Video, and export the film to camera roll.

Extensions

Students could create and record the soundtracks/sound effects for their films. Groups could collaborate to produce a film made up of a variety of scenes/settings as part of a larger project. Students could combine stop-motion with hand-drawn animation with an app such as Do Ink Animation.

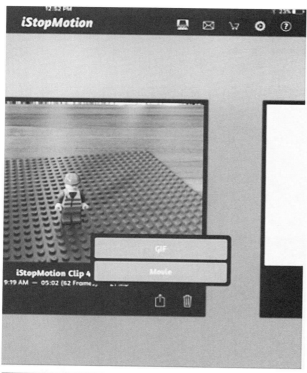

Figure 6-121

Figure 6-122

Figure 6-123

14. Create Your Own Comic

In this project, students will plan and create their own comic book or graphic novel.

Setup

For this project, we'll use some paper and pencil planning tools and will then use technology to help us make the vision a reality.

Materials

- Plot diagram
- Comic Planning Sheets
- Costumes
- Props
- Background images

Technology

- Green Screen by Do Ink app
- Book Creator app
- Magic Eraser app
- Green screen
- Tripod holder
- iPad

Instructions for Students

1. Complete the plot diagram via ReadWriteThink (www.readwritethink.org/files/resources/interactives/plot-diagram/) to develop a sense of story sequencing and plot elements (Figure 6-124).

2. After you have developed your plan, use one of the comic layouts to choose the photographs you want to use to tell your story by sketching the perspective and text of each "cell."

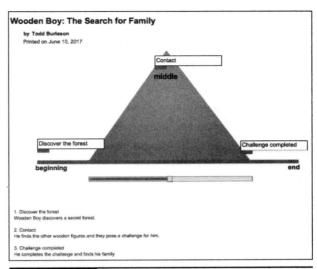

Figure 6-124

3. Next, photograph yourself and other characters in front of the green screen.

4. Search for background images that help you to communicate your story.

5. Import each background into the green screen, placing the background image as the bottom layer (Figure 6-125).

Figure 6-125

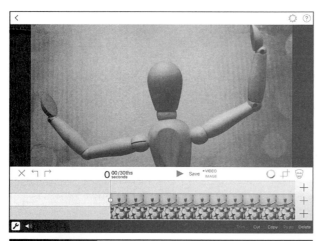

Figure 6-126

Figure 6-128

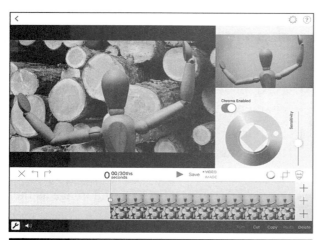

Figure 6-127

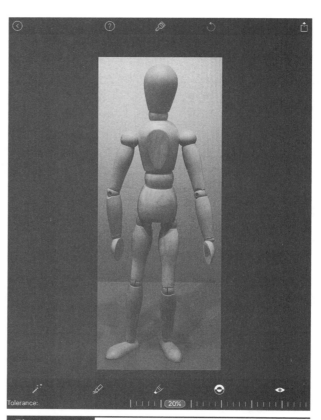

Figure 6-129

6. Insert the green screen image as the top layer (Figure 6-126).

7. Adjust the Chroma key wheel to the appropriate shade of green (Figure 6-127).

8. Slide the Video/Image switch to Image, and save the image to the camera roll.

9. I decided I wanted to have multiple versions of one of the images. In order to do this, I used the Magic Eraser app to extract the background (Figure 6-128).

10. After opening the app, you can either take a photo or select a photo from your camera roll; I chose the image I wanted to duplicate (Figure 6-129).

11. I began removing the background by selecting the Magic Wand tool and then tapping on the green background color (Figure 6-130).

12. The Magic Wand tool does a very good job, but I needed to use the eraser to remove smaller sections. The app allows you to use the two-finger pinch to zoom into these tiny areas (Figure 6-131).

13. If you erase too much, you can simply click on the Restore pen, and when you draw over the area that was inadvertently erased, it will reappear (Figure 6-132).

14. You can tap the inverse to see exactly what you have selected and decide whether you need to do some more cleaning up before saving.

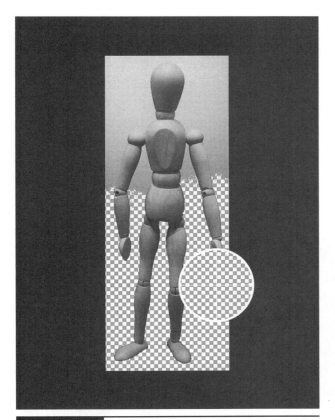

Figure 6-130

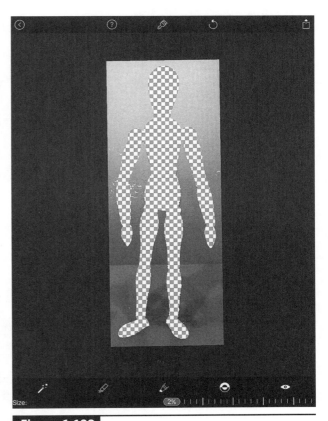

Figure 6-132

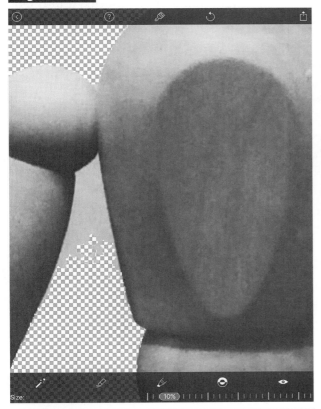

Figure 6-131

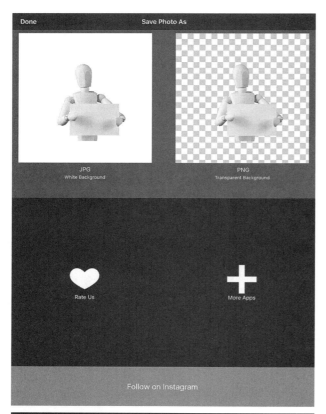

Figure 6-133

15. When you are happy with your selection, save it (Figure 6-133).

16. You will be asked whether you would like to save the selection as a JPG file with a white background or a PNG file with a transparent background. By selecting a transparent background, you can layer many versions of the same image.

17. I decided that I wanted to create an image that looked like a large group of clones was standing in a dark forest. To do this, I used Do Ink's ability to create three distinct layers.

18. I imported the background image.

19. Next, I imported the PNG file I had just created in the Magic Eraser app (Figure 6-134).

20. I imported a second version of the PNG file and used the two-finger pinch to make it smaller. I next saved these three layers as a still frame and then reimported it into Do Ink, making it the bottom layer again.

Figure 6-134

21. Then I simply continued doing this process until I was satisfied with the number of duplicates I had created (Figure 6-135).

22. I decided to add a filter in the Photos app to make the figures look more like the dark background (Figure 6-136).

23. When you have created all your images, open Book Creator (Figure 6-137).

24. Select New Book from the top-left corner.

25. Choose a book shape. I chose Landscape (Figure 6-138).

26. Create your cover by importing photographs and adding text; you can also add color to the full page (Figure 6-139).

Figure 6-135

Figure 6-137

Figure 6-136

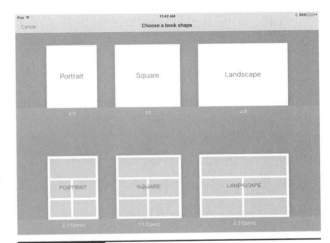

Figure 6-138

Figure 6-139

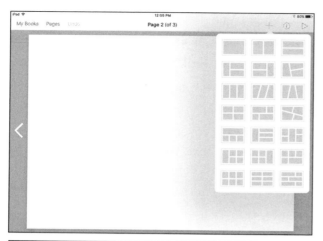

Figure 6-140

Figure 6-141

27. Select the layout of your comic book panels (Figure 6-140).

28. Import your images, and add stickers, text, captions, thought bubbles, and so on (Figure 6-141).

29. When you are pleased with the pages of your book, you can preview the layout of your pages before exporting to the camera roll (Figure 6-142).

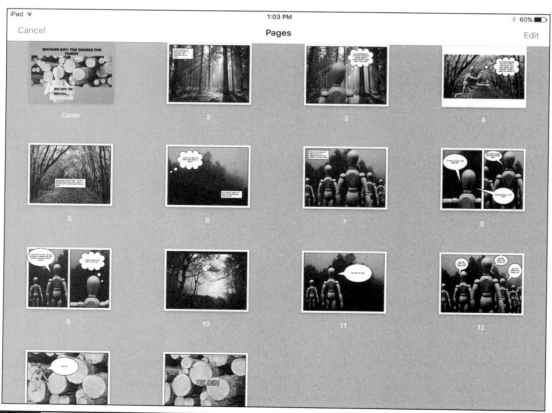

Figure 6-142

30. You can click Read to Me, and an automated voice will read all the words in your book to you; it's a bit robotic, but it is still neat to hear your words read aloud (Figure 6-143).

31. By clicking on the Adjustment tab, you can determine how the book appears to your readers and tweak a few additional elements (Figure 6-144).

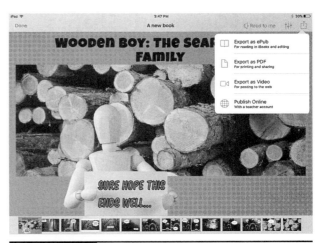

Figure 6-145

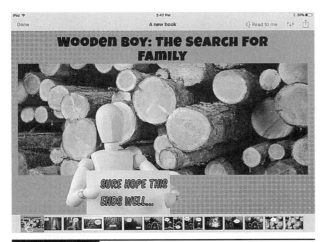

Figure 6-143

32. When you are pleased with your book, tap the Share icon. You will have four ways to export your book (Figure 6-145).

Extensions

This example uses only still images, but with the Book Creator app, it is possible to integrate video segments into the comic cells as well.

Figure 6-144

15. Find Geometric Shapes in Architecture

This project comes to us from third grade teacher Patrick Johnson from Maple Grove Public School in Ontario, Canada. He used it as a way to have his students use their "math eyes to find 3D shapes in architecture."

Setup

Students use wooden models of basic geometric solids and images of famous architectural landmarks to find examples.

Materials

- Wooden examples of basic geometric shapes
- Background images

Technology

- iPad
- Green screen
- iPad stand
- Green Screen by Do Ink app

Instructions for Students

1. Explore images on the following website to find examples of how basic geometric shapes are present in famous pieces of architectural design at https://photography .tutsplus.com/articles/25-stupendous -examples-of-architecture-photography --photo-7566.

2. After locating the basic geometric shapes, download the image to the camera roll (long tap on an iPad, and choose to save the image to the camera roll) (Figure 6-146).

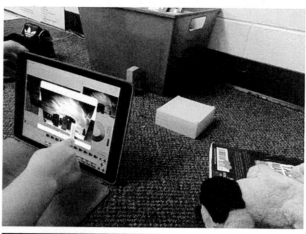

Figure 6-146

3. Next, import the photograph of the architectural example, making it the bottom layer in the Green Screen by Do Ink app.

4. Place a wooden geometric solid in front of the green screen (in this case, the students used plastic library totes for a green screen!) (Figure 6-147).

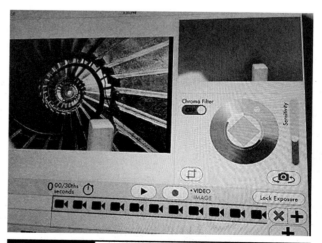

Figure 6-147

5. After positioning the geometric solid where you want it, take the photograph right within the Do Ink app (Figure 6-148).

6. Use the color wheel to remove the green screen (Figure 6-149).

7. After using the two-finger pinch/zoom, save the composition by sliding the Image/Video slider to Image and saving the image to the camera roll.

Figure 6-149

Figure 6-148

Extensions

Students could add a third layer to this project and record a video of themselves interacting with the geometric solid and the downloaded photograph. They could pinch and zoom themselves to make it appear as if they are tiny and are walking around both the geometric solid and the architectural example.

16. Pic Forward Global Green Screen Challenge

This project comes to us from Larissa Aradj, an educator from the Ecole Lord Lansdowne Public School in Ontario, Canada. In this project, students will participate in a global green screen challenge that fosters creativity, connection, and collaboration. The Pic Forward project gives students a chance to travel the world without ever leaving the classroom. It is a fantastic way to develop potential relationships and collaboration with classes in other provinces, states, and countries around the world. This project is like a twenty-first-century version of the famous Jeff Brown book, *Flat Stanley*. This book, published in 1964, has helped students to connect with others all over the world (Figures 6-150 and 6-151).

Figure 6-150

Figure 6-151

Pic Forward is a global green screen challenge for students, teachers, and schools. The idea is to have people all around the world edit their monthly green screen photographs and then to "Pic Forward" or "pass them on" via social media to other classes so that they can participate too. Fourth and fifth grade students lead this project to spread creativity around the globe.

Materials

■ Creativity!

Technology

■ Green screen

■ Tripod stand

■ iPad

■ Green Screen by Do Ink app for iPad

Using the template, the entire process can be completed within the Google Suite on a Chromebook or any other tablet as well.

Instructions for Students

1. Visit the Pic Forward Global Green Screen Challenge webpage to download the current green screen photos from the students (bit.ly/PicForward).

2. Download the image you want to Pic Forward (Figure 6-152).

3. Choose a background in which you would like to place the student(s). Remember to use images that are "labeled for reuse."

4. Import the background image to the green screen, making it the bottom layer. You will need to use the two-finger pinch to resize it (Figure 6-153).

Figure 6-152

Figure 6-153

5. Import the green screen photo you are going to Pic Forward, making it the top layer. You will need to resize it using the two-finger pinch/zoom (Figure 6-154).

6. Use the color wheel to adjust the Chroma key color to match the green of your background; adjust the sensitivity to dial in the image to your liking (Figure 6-155).

7. Use the two-finger pinch and zoom to adjust the size and location of the pic Forward image (Figure 6-156).

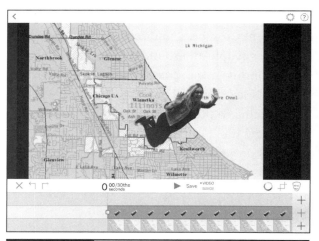

Figure 6-156

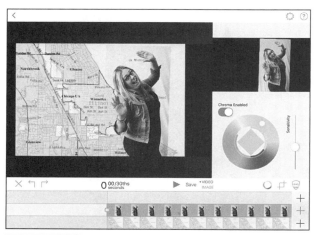

Figure 6-154

8. When you are pleased with the two layers, slide the Image/Video tab to Image, and save the image to the camera roll (Figure 6-157).

9. Share online via @PicForward, #PicForward, or the website (bit.ly/ PicForward) (Figure 6-158).

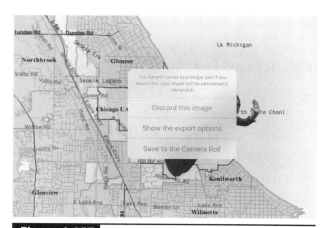

Figure 6-157

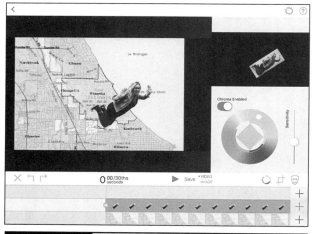

Figure 6-155

Figure 6-158

Extensions

This is just the beginning. Students can "tag" another school to encourage it to participate, or they could take their own green screen photographs and send them back to the students in Ontario to give them the chance

to participate! Students could also use Do Ink Animation to animate the green-screened students. They could use an app such as Google Drawing to add text or speech bubbles.

Tip: Larissa has now created a template that allows you to edit, search for backgrounds, and more right within Google Drawings. This is especially helpful for students who are using Chromebooks or laptops.

1. Visit the website https://sites.google.com/site/aradjpicforward/home (Figure 6-159).

2. Click on the template, which will ask you to make a copy within your Google account.

3. Find a background:
   Insert > Image > Search

4. Slide in the students, and add text bubbles. Be creative!

5. Download your finished image:
   File > Download as ... > JPEG (.jpg)

6. Share with us at @PicForward, #PicForward, or bit.ly/PicForward.

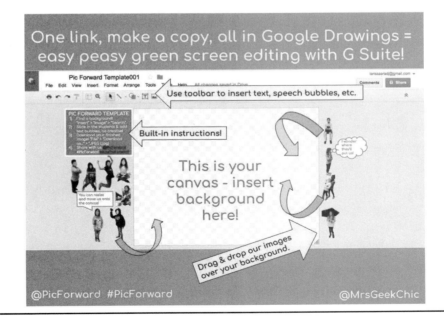

Figure 6-159

17. Storytelling Project

This project comes from first grade teacher Chloe Milligan of the Adams Elementary School in the Santa Barbara Unified Schools District. In this project, Chloe had staff members take a variety of action shots in front of a green screen with all sorts of costumes and props. Students then took those images, processed them in green screen editing software to change the backgrounds, and then wrote creative pieces with their teachers as the characters.

Setup

Chloe used this project to help her students learn the ins and outs of the software before getting in front of the camera. By demonstrating how easily the process of removing the green screen was and having her students explore the process of locating and then using unique backgrounds, she was preparing them for using the green screen software in more advanced ways.

Materials

- Props
- Costumes
- Planning sheet

Technology

- Green screen
- iPad
- Tripod
- Green Screen by Do Ink app

Instructions for Students

1. Have staff members choose their costumes and props and then pose in front of the green screen in a variety of fun poses (Figure 6-160).

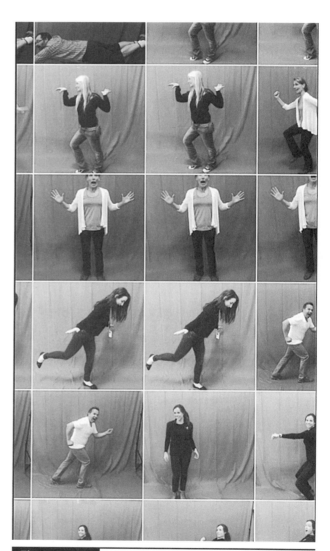

Figure 6-160

Figure 6-161

Figure 6-162

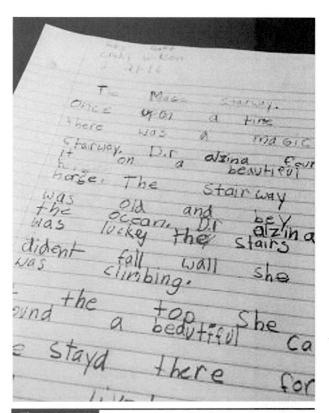

Figure 6-163

2. Capture the images using the camera roll (Figure 6-161).

3. Save the images to a shared folder that you can access.

4. Complete a planning sheet to help you organize your ideas for plot, backgrounds, setting, characters, and so on.

5. Search for backgrounds into which you would like to place your teachers.

6. Place your background image as the bottom layer of the Green Screen by Do Ink app.

7. Place your green screen image as the top layer.

8. Use the color wheel to remove the green screen (Figure 6-162).

9. Slide the Video/Image slider to Image.

10. Save your image to the camera roll.

11. Import your image into the word processing software of your choice and develop your creative writing piece (Figure 6-163).

12. Share with classmates.

Extensions

Students could photograph one another and use themselves as the characters in their own stories. Students could develop a persona, such as a superhero, and develop the story into a comic book (see Projects 1 and 14).

18. Book Talks

Karen Hopis is a library media specialist from Erie, CO. She had a conversation with her content educators about how to enhance the culture of reading within their school. They asked the question: "How do we help students see their teachers as readers?" Her solution was to use a green screen and Quick Response (QR) codes to literally show the teachers in the school reading and making book recommendations to their students (Figure 6-164).

Setup

Karen had each staff member pick a piece of middle-grade fiction that they would recommend to their students. She then created a Google Slide Deck that walked the teachers step by step through the process of creating a book recommendation (Figure 6-165).

Materials

- Books
- Background images

Technology

- Green screen
- Green Screen by Do Ink app
- iPad
- Tripod
- QR code generator

Figure 6-164

Figure 6-165

Instructions for Students

1. Have teachers choose a book they would like to recommend.

2. Have each teacher search for an appropriate image for the background.

3. Use the Camera app to take a photograph of the teacher with the book in his or her hand in front of the green screen background. (Figure 6-166).

4. Import the background image into the Green Screen by Do Ink app as the bottom layer (Figure 6-167).

Figure 6-168

5. Import the teacher-selected photograph in front of the green screen, making it the top layer (Figure 6-168).

6. Use the Eraser tool to remove the extra background area beyond the green screen (Figure 6-169).

Figure 6-166

Figure 6-167

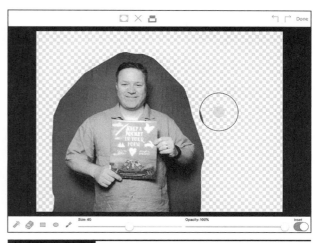

Figure 6-169

Figure 6-170

Figure 6-171

7. Adjust the color wheel to remove the green background (Figure 6-170).

8. Use the two-finger pinch and zoom maneuver to resize the photo of the teacher.

9. Have the teacher either create a video "book talk," read a favorite section of the book, or whatever else he or she may want to share with students. Save this in Google Drive.

10. Create a link to this content.

11. Use a QR code generator to create a QR code with the link to this content.

12. Download the QR code.

13. Import this QR code, and make it the top layer in Do Ink (Figure 6-171).

14. When you are happy with the composition, slide the Image/Video slider to Image, and save the image to the camera roll (Figure 6-172).

Figure 6-172

Figure 6-173

15. Print these images out, and display them with copies of the book (Figure 6-173).

Extensions

Karen encouraged her teachers to consider making a video book recommendation. These can be recorded using the Camera app and then imported into Do Ink as a video layer. Saving these online and using a QR code creator, students could access these video recommendations using a QR Scanner app on their phone or tablet.

19. Interview with a President

In this project, students will imagine what it would be like to interview a president, and then they actually will conduct an interview with a president from the past.

Setup

Imagine if you could ask any president in the history of the United States five questions. What would they be? How do you think he would answer? In this project, students get to do just that.

Materials

- Planning sheet

Technology

- iPad
- iPad tripod stand

- Green screen
- Green Screen by Do Ink app
- Puppet Pals 2 iPad app

Instructions for Students

1. Work in teams to research your president and come up with five questions you would like to ask him.

2. You will also need to research the president's responses as well. While it might not be possible to know exactly his response, the responses should be based on facts and events that had a major impact on his life.

3. When you have your questions and answers written out and have practiced them several times, open the Puppet Pals 2 app (Figure 6-174).

Figure 6-174

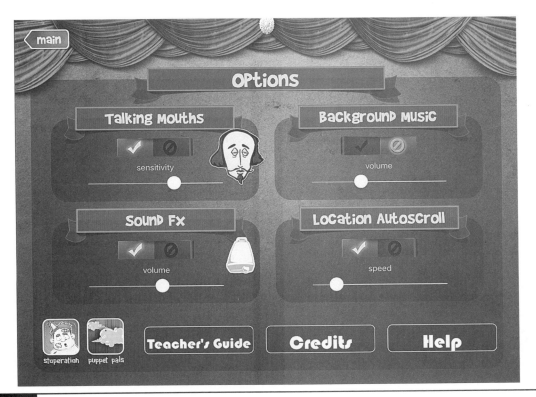

Figure 6-175

4. Tap on the Options tab. For this project, I turned off the background music (Figure 6-175).

5. Tap the arrow to get back to the main screen; then tap the Play tab.

6. You will next choose from a list of locations. For this project, we are going to use a photograph of a green screen we've placed in our camera roll so that we can place the president wherever we want him to be (Figure 6-176).

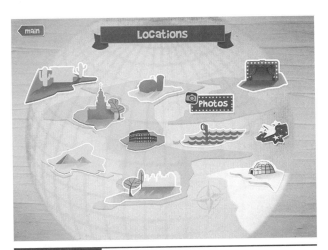

Figure 6-176

7. After selecting the green background, your "stage" will be a rectangle of solid Chroma key green (Figures 6-177 and 6-178).

8. Next, tap the Character icon to select your actor. In the free version of the app, you are limited to a few characters. For this project, we are going to use President Abraham Lincoln. Tap him to select (Figure 6-179).

9. For this project, the president is going to stand on your shoulder, so he will not be moving around. If you tap and hold him, you can change the direction in which he is looking. You can also adjust his arms and legs by pressing and moving them. When you are ready, tap the Record button (Figure 6-180).

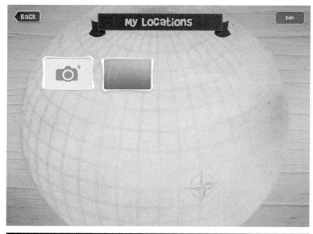

Figure 6-177

Figure 6-179

Figure 6-178

Figure 6-180

10. When you want the president's mouth to move, you need to press and hold his character as he "speaks." Don't worry, you won't actually be using the audio track; you are just having him move his mouth. You'll record him speaking later in the project.

11. When you are done, press the Stop button. You will then be asked to name the recording (Figure 6-181).

12. After saving, you will export the recording to the camera roll; I saved the high-quality version (Figure 6-182).

Figure 6-183

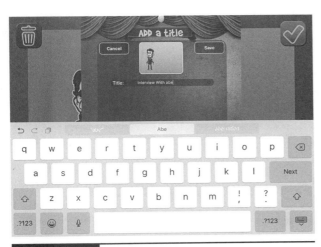

Figure 6-181

Figure 6-182

13. Now you are ready to import your video into the green screen app.

14. Create a new project.

15. Select a solid-color image as the background, making it the bottom layer (Figure 6-183).

16. Next, import the recording of the president, making it the top layer.

17. For the middle layer, select the Camera icon, and select the rear-facing camera so that you can see where the president will appear in the frame.

18. Use the two-finger pinch and zoom maneuver to adjust the size and location of the president.

19. When you have him in just the right place, begin recording. You will see his mouth moving when he is "talking." Try to turn your head toward the president when you are "listening."

20. Your partner will need to read the answers in the voice of the president while you are listening.

21. When you have finished asking the questions and listening to his answers, tap the Stop button.

22. Tap to adjust the audio (Figure 6-184).

23. Slide the prerecorded audio slider all the way down.

24. Tap Done (Figure 6-185).

25. Save the project to the camera roll.

26. View your interview!

Extensions

There can be a total of eight actors on the "stage" at one time. Students could design and photograph their own backgrounds. The edu version of the app includes all the actors, locations, and so on. The vast number of characters is impressive. The app includes background music and sound effects as well. Students can even put themselves in the puppet show by adding their faces to the puppets.

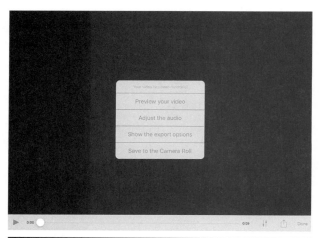

Figure 6-184

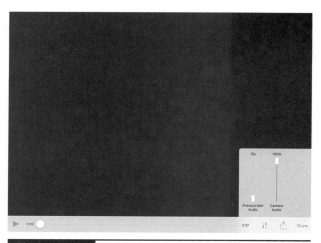

Figure 6-185

20. TeleStory

TeleStory is a video creation app that allows kids to write, direct, and star in their own television shows. The app has facial recognition and sticker-like costumes that enhance each of the broadcast types. Students can choose from "The News," "The Band," "Eye Spy," and "Star Adventure" (Figure 6-186). They can then create whole TV shows by choosing segments around a common theme. Each segment includes some fun features such as cue cards, multiple cameras, and special video and sound effects. Students can then save their broadcasts to their camera rolls and even to the web in a parent-monitored and -approved channel.

In this project, students will create their own newscast, complete with reports on weather, sports, and more. The segments will be filmed in front of a green screen so that the background can be enhanced in the Green Screen by Do Ink app afterward.

Setup

Students love creating and sharing their own news broadcasts. TeleStory makes this fun. Students should storyboard each of the segments of the broadcast, including cue cards.

Figure 6-186

Materials

- Storyboard
- Planning sheet

Technology

- iPad
- Tripod
- iPad stand

- Green Screen by Do Ink app
- TeleStory app

Instructions for Students

1. Decide who is going to record each segment of your newscast and then sketch it out on the storyboard, being sure to include any text you would like on your cue cards (Figure 6-187).

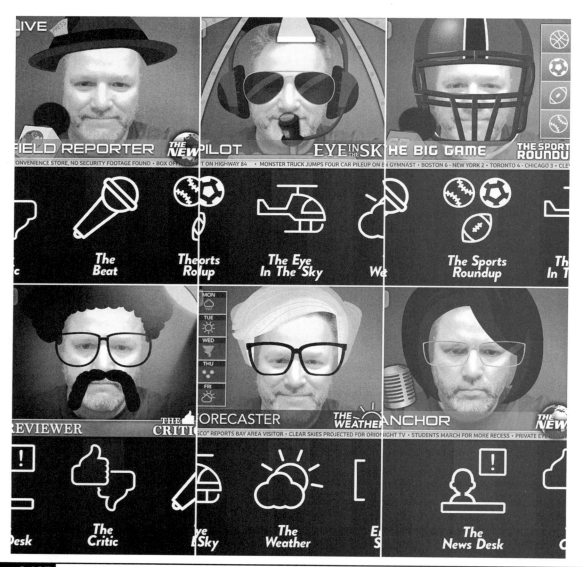

Figure 6-187

Figure 6-188

2. After you have planned each of your segments, add your text to the cue cards; you are limited to one card per segment (Figure 6-188).

3. When you are ready, tap the Start button to begin recording. Each segment has two "cameras" that give a different perspective. There are also two sound or action effects that correspond to the theme as well. You can tap these while recording your segments to enhance the overall impact (Figure 6-189).

4. When you are done, press the Stop button (Figure 6-190).

Figure 6-190

Figure 6-189

5. Now you may preview the video, rerecord it, or save it (Figure 6-191).

6. Place this segment in the timeline for the complete broadcast. Then you can tap the Plus button to add another segment or the Check button to go ahead and produce the entire broadcast (Figure 6-192).

Figure 6-191

Figure 6-193

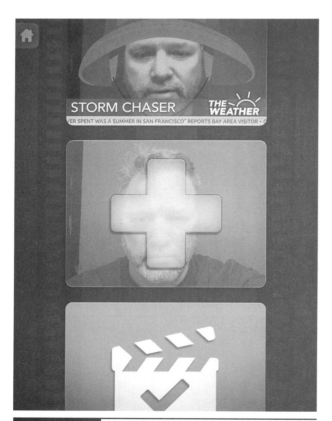

Figure 6-192

7. When the entire broadcast is complete, you will be asked to add the details of the broadcast to the "clapper" (Figure 6-193).

8. Save the broadcast to the camera roll.

9. Next, open the Green Screen by Do Ink app to create a new project.

10. Choose a background image with which you would like to replace the green screen image for each of the segments. Load that image, making it the bottom layer. I loaded a solid black image to the timeline so that the opening credits were not affected by the green screen. Click on the middle layer, and drag the red "handle" to adjust the length of each of the background images to match the effect you are trying to create (Figure 6-194).

Figure 6-194

11. Adjust the Chroma key color by sliding the color wheel to the appropriate color and tweaking the sensitivity slider (Figure 6-195).

Figure 6-195

12. When each segment has been adjusted, preview the broadcast. When pleased with the overall effect, slide the Video/Image slider to Video, and save the video to the camera roll.

Extensions

Obviously, this is just one of the four built-in templates from which students can choose. Each one offers a multitude of options and features that are sure to encourage creativity and collaboration. Each template also includes an "empty shot" that allows students to choose from all the costumes, further enhancing the creative potential.

21. ThingLink

ThingLink is a tool that allows users to enhance images and videos with embedded content such as notes, audio, photo, video, and other multimedia content. Educators can sign up for a free account.

In this project, students will insert themselves into a bird's-eye-view map of their school's library, classroom, or any other interesting location. Then they will research various elements of the space and the work done there and add multimedia content so that virtual "visitors" can grow to know more about the learning environment.

Setup

We have many visitors to our resource center, and we are very eager to share our passions with them. Some visitors, though, cannot make it in person. For this reason, we wanted to create a virtual tour of our Library and makerspace. Students will do the same for their school.

Materials

- Map of your space. This was created in Google Drawings.

Technology

- iPad
- Green screen
- Tripod stand
- Green Screen by Do Ink app
- Book Creator app

Instructions for Students

1. Assist your teacher in creating a bird's-eye map of your learning space using the Google Drawings tools. Figure 6-196 shows our map.

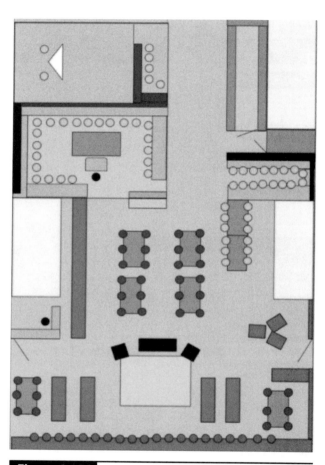

Figure 6-196

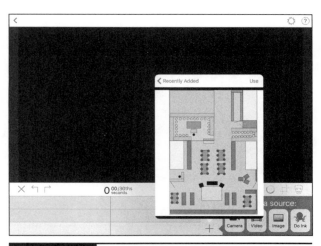

Figure 6-197

2. You should add an image of a student to the map to visually welcome visitors and help them navigate the ThingLink map. To do this import the map into the Green Screen by Do Ink app. Figure 6-197 show our result.

3. In our project, we needed a little more "room" to place the student, so we also imported a solid-white background image to allow us to move the map off to the side. You should do the same. Make this solid-white background image the bottom layer

and your map the middle layer Figure 6-198 shows our result.

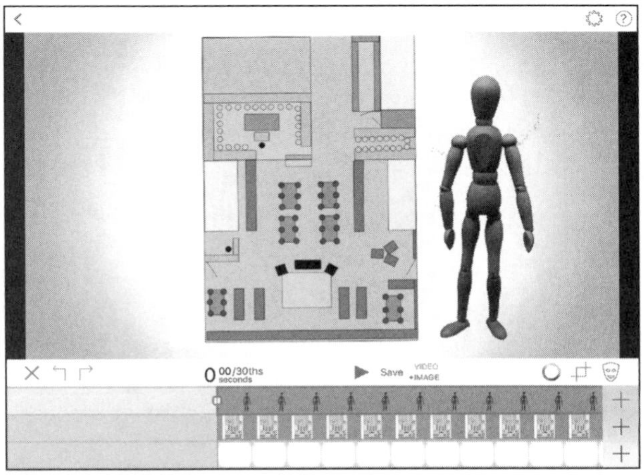

4. Lastly, add the image of the student standing in front of the green screen.

5. After adjusting and removing the background, slide the Video/Image slider to Image, and save the image to the camera roll. Figure 6-199 shows our result.

6. We wanted to make the image a little livelier, so we decided to add a speech bubble. There are a variety of ways to do this and we

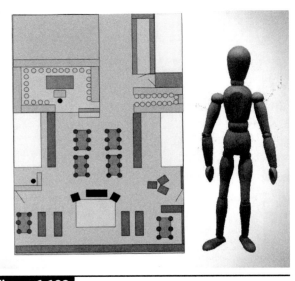

Figure 6-199

Figure 6-198

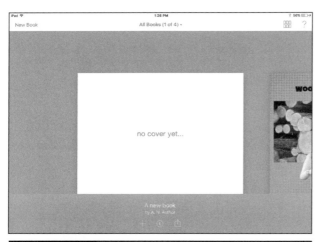

Figure 6-200

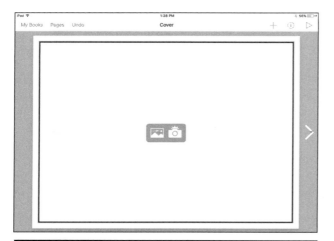

Figure 6-201

chose to use Book Creator. You should do this as well.

7. To do this, create a new book in Book Creator, and chose the landscape comic template. Figures 6-200 and 6-201 show how we did this.

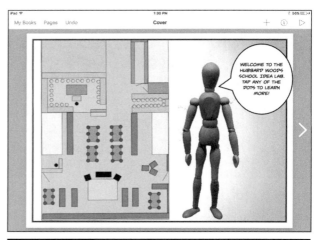

Figure 6-202

8. We added a speech bubble to welcome visitors to the map (Figure 6-202). You should do the same.

9. Save this composition to the camera roll. Figure 6-203 shows our result.

10. Brainstorm about how each of the different "areas" of your space could be highlighted. Search online for articles that may have appeared in your local papers. Visit your school's YouTube page and select images

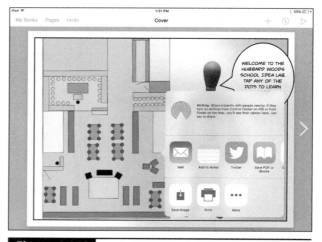

Figure 6-203

or links that you think will be helpful for virtual visitors to your center.

11. Share these resources on a shared Google document.

12. In small groups, save these links and add them to your map in ThingLink.

13. Simply copy the URL of the item you want to share and tap on the map and paste the link. Figure 6-204 shows our result.

14. You also could add text or even video elements right within the app.

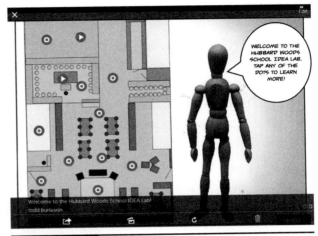

Figure 6-205

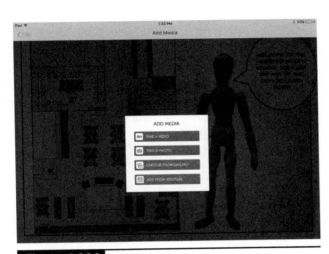

Figure 6-204

15. When you are pleased with all the links, save the composition to ThingLink. ThingLink saves these to your account, and you can then copy and paste the link in a variety of ways, including embedding on a website (Figure 6-205).

Extensions

Students could use ThingLink and green screens to augment research projects. Students could create a "living timeline" and film various segments of a famous person's life in front of a green screen and then add in background elements or features. By combining them with ThingLink, the multimedia interactive project can be self-contained and easily shared.

22. ChatterPix Kids

ChatterPix Kids is a free app that allows you to make anything "talk." Simply take a photograph, draw a line over the mouth, record what you want to say, and share. The app allows users to add fun stickers and other special effects and then share in a variety of formats.

In this project, students will use the green screen to place a photograph of themselves in front of a winter-themed background. They will then use the ChatterPix Kids app to make their photographs speak.

Setup

After demonstrating a few examples of ChatterPix Kids, students will create their own.

Materials

- Creativity!

Technology

- iPad
- Green Screen by Do Ink app
- ChatterPix Kids app
- Tripod stand
- Green screen

Instructions for Students

1. Have a partner take a photograph of you in front of a green screen; save the photo to the camera roll (Figure 6-206).

2. Choose an image you would like as your background. Create a new project in the Green Screen by Do Ink app, and make this background the bottom layer (Figure 6-207).

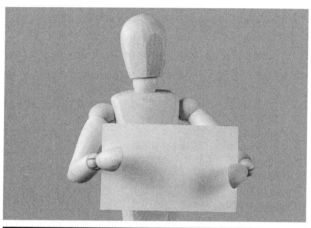
Figure 6-206

Figure 6-207

3. Adjust the color wheel to remove the background.

4. When satisfied with the composition, slide the Video/Image slider to Image, and save composition to the camera roll (Figure 6-208).

5. Open the ChatterPix Kids app (Figure 6-209).

6. Tap the Photo icon, and then choose the image from your camera roll (Figure 6-210).

7. Select the image you just created (Figure 6-211).

Figure 6-208

Figure 6-210

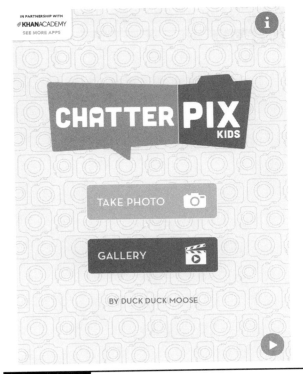

Figure 6-209

Figure 6-211

8. Use your finger to draw a line where you would like your image to "speak." You can record up to 30 seconds of audio.

9. Tap the Microphone icon to begin recording. You can listen to your recording by tapping the green Play icon. You can redo the recording by tapping the Microphone icon again. When you are ready to move on, tap the yellow Next arrow (Figure 6-212).

10. You will then have then have access to a variety of fun options: filters, stickers, frames, and text (Figure 6-213).

Figure 6-213

Figure 6-212

11. Since we wanted to have a winter theme, I selected some fun winter stickers (Figure 6-214).

12. The last piece I added was text for the sign the student is holding (Figure 6-215).

13. When all the elements you want to add are in place, save the composition to the camera roll. Figure 6-216 shows my result.

Extensions

Students could insert these videos into other projects via the camera roll. Imagine a "talking" comic book or famous work of art. Biography projects could "come to life" as students voice their subject. Animal research projects can speak for themselves.

Figure 6-214

Figure 6-215

Figure 6-216

23. Autobiography of Your Future Self

Tellagami is a free app (with in-app purchases) that allows you to create a virtual avatar for which you can record your own voice and speak text that you enter for it. The app allows you to customize the look, mood, and clothing and even set how your avatar appears. These fun short videos can then be shared quickly and easily.

In this project, students will imagine themselves 10 years into the future. They will ask themselves five questions and then answer as their future self, thanks to Tellagami. The "gami" will be combined with a video of the student asking the questions.

Setup

This project is a great way for students to think about their future and how they hope it will turn out. They will have a chance to "interact" with the future selves they hope they will become.

Materials

- Brainstorming materials

Technology

- iPad
- Green screen
- Tripod
- Green Screen by Do Ink app
- Tellagami app
- Google Slides app

Instructions for Students

1. Brainstorm where you would like to be in 10 years. Where do you hope to live and work? What do you hope to have accomplished? Decide on five questions you would like to "ask" your future self.
2. Use the Google Slides app to create a title slide and "cue cards" that will appear in the movie (Figure 6-217).
3. After creating each of the "slides," take screen shots of each and save them to the camera roll.
4. Open the Tellagami App (Figure 6-218).

Future Autobiography
An Interview with My Future Self

Figure 6-217

Figure 6-218

Figure 6-219

5. Tap Create (Figure 6-219).

6. Tap the Character tab, and select the gender, clothing, hair color, skin tone, and emotion of your avatar. (Figure 6-220).

Figure 6-220

7. Take a photo of your green screen or download a solid-green image from the web. Select this as the background. This will allow you to layer the "gami" with video of you in front of the green screen (Figure 6-221).

8. Next you either need to record your own voice as the "gami" or type the text that you would like your "gami" to speak. For this project, record your voice. (You can record a maximum of 30 seconds in the free version of the app.) To create the illusion that your "gami" is listening to you ask the question, record the first 10 to 15 seconds of silence and the last 15 seconds answering your question (Figure 6-222).

Figure 6-221

Figure 6-222

Figure 6-223

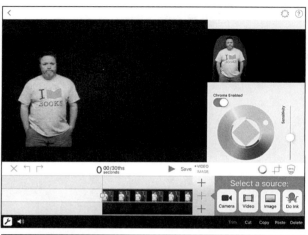

Figure 6-224

9. You may preview your recording, and when you are satisfied with it, tap the Save icon (Figure 6-223).

10. The "gami" will save the recording to the camera roll. Follow this same process for each of the five questions.

11. Next, record a video of you in front of the green screen with the iPad camera asking each of the five questions. (Remember that you will need to be silent for 10 to 15 seconds after asking your question so that your "gami" can answer. It will be easier if you have your partner stop and save each individual answer and response as a separate movie).

12. Next, combine both you and your "future you" together in the green screen. Open the Do Ink app, and create a new project.

13. Download a solid black background, making it the bottom layer.

14. Import your "current self" asking the first question, making yourself the middle layer. Tap and adjust the color of the green screen to erase the background (Figure 6-224).

15. Use the two-finger pinch/zoom maneuver to reposition yourself on the left of the frame.

16. Import the "gami" answering the first question, making it the top layer (Figure 6-225). Adjust the color wheel to remove the green screen (Figure 6-226).

Figure 6-225

Figure 6-226

Figure 6-227

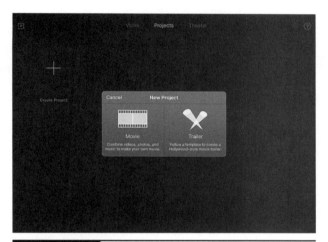

Figure 6-228

17. Use the two-finger pinch/zoom maneuver to position your "gami" on the right side of the screen.

18. "Gamis" have a blue "tag" at the end with the Tellagami logo. To get rid of this, tap the red "handle" to adjust the length of both recordings and line them up as needed.

19. Preview the timing of both recordings.

20. When you are satisfied, export the recording to the camera roll (Figure 6-227).

21. Save the project to the camera roll.

22. Do this for each of the five questions.

23. Open the iMovie iPad app, and create a new "movie" project (Figure 6-228).

24. The media browser will open up. Select each of the "cue cards" and the combined video segments, and then tap Create Movie (Figure 6-229).

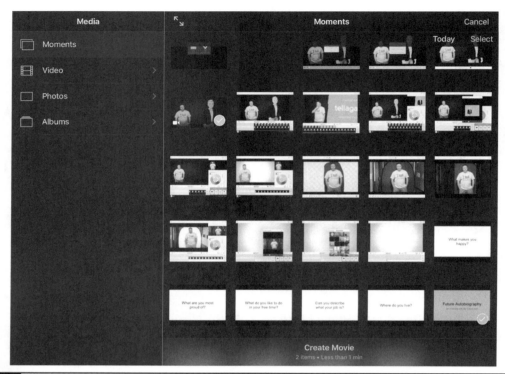

Figure 6-229

25. You can rearrange the order of the cue cards and videos by "long touching" each of the items and then dropping them in the timeline where you would like them to be.

26. When you have placed all the items in the timeline in the order you would like them to be and have previewed the project, tap done (Figure 6-230).

27. Name your project, and then tap the Share icon to save the project to your camera roll (Figure 6-231).

Extensions

Students could create a variety of "gamis" and layer them in the green screen. Students could also create their own backgrounds for their "gamis."

Figure 6-230

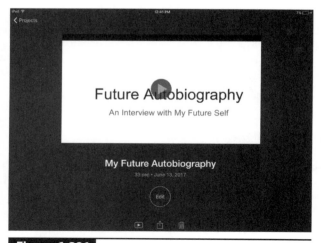

Figure 6-231

24. Shadow Dancing

This project comes from an amazingly creative art teacher from Arlington Heights, IL. Tricia Fuglestad pushes apps beyond their limits and discovers fun and unique ways to combine them with other apps.

In this project, students will learn to apply a green screen technique using the Green Screen by Do Ink app that creates a silhouette, copies it, and changes the transparency to create a shadow and then layer it over a background (Figure 6-232).

Setup

The project connects dancing, music, and visual art while teaching about shadows, transparency, silhouettes, complementary colors, digital layering, and movement.

Materials

■ Creativity!

Figure 6-232

Technology

■ iPad
■ iPad tripod stand
■ Green screen
■ Green Screen by Do Ink app
■ iMovie app
■ Background image

Instructions for Students

1. Have a partner take a green screen video of you dancing in front of the green screen with the Camera app (Figure 6-233).

Figure 6-233

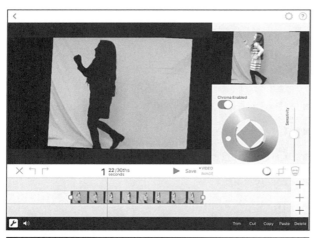

Figure 6-234

Figure 6-236

2. Import that video into the green screen app, making it the middle layer (Figure 6-234).

3. Create a silhouette by clicking on the color wheel button, and choose a spot on the other side of the color wheel for the Chroma key effect. The complementary color (choose red instead of green) makes the subject a silhouette.

4. Adjust the sensitivity bar, and choose the Crop button to clean up your video.

5. Slide the Video/Image button to Video, and save the video to the camera roll.

6. Download the background (stage) image (Figure 6-235).

7. Create a new project in the Green Screen by Do Ink app, making the stage image the bottom layer (Figure 6-236).

8. Tap the "plus," and add your video of the solid-black student dancing as the middle layer.

9. Use the two-finger pinch/zoom maneuver to position the dancing figure on the stage.

10. Tap on the middle layer to activate it, and tap Copy at the bottom of the screen.

11. Touch the top layer, and then tap Paste to insert a duplicate version of the student dancing.

12. Tap the top layer and adjust the sensitivity slider until the top layer becomes almost transparent.

13. Use the two-finger pinch/zoom maneuver to reposition the shadow off to the right side of the silhouette.

14. Delete the audio from both video layers; the audio will be combined later in iMovie when multiple students are combined together.

Extensions

Tricia's website (http://drydenart.weebly.com/) is full of incredible explorations of all aspects of green screens and the visual arts. She explores both still and moving silhouettes with her students and truly pushes the limits of many apps.

Figure 6-235

Index

References to figures are in italics.